Computerized Manufacturing Process Planning Systems

Computerized Manufacturing Process Planning Systems

Hong-Chao Zhang

Assistant Professor
Department of Industrial Engineering
Texas Tech University
Lubbock
Texas
USA

and

Leo Alting

Institute of Manufacturing Engineering
Technical University of Denmark
Lyngby
Denmark

CHAPMAN & HALL

London · Glasgow · New York · Tokyo · Melbourne · Madras

Published by Chapman & Hall, 2–6 Boundary Row, London SE1 8HN

Chapman & Hall, 2–6 Boundary Row, London SE1 8HN, UK

Blackie Academic & Professional, Wester Cleddens Road, Bishopbriggs, Glasgow G64 2NZ, UK

Chapman & Hall Inc., One Penn Plaza, 41st Floor, New York NY10119, USA

Chapman & Hall Japan, Thomson Publishing Japan, Hirakawacho Nemoto Building, 6F, 1-7-11 Hirakawa-cho, Chiyoda-ku, Tokyo 102, Japan

Chapman & Hall Australia, Thomas Nelson Australia, 102 Dodds Street, South Melbourne, Victoria 3205, Australia

Chapman & Hall India, R. Seshadri, 32 Second Main Road, CIT East, Madras 600 035, India

First edition 1994

© 1994 Hong-Chao Zhang and Leo Alting

Typeset in 10/12 Palatino by Best-set Typesetter Ltd., Hong Kong
Printed in Great Britain by Clays Ltd, Bungay, Suffolk

ISBN   0 412 41300 0

A catalogue record for this book is available from the British Library

Library of Congress Catalog Card Number available

♾ Printed on permanent acid-free text paper, manufactured in accordance with the proposed ANSI/NISO Z 39.48-199X and ANSI Z 39.48-1984

Contents

Preface

Over the past three decades or so, great advances have been made in the development of automated process planning systems. While the goal of the research remains the same, the content and emphases have undergone tremendous changes during this period. Although significant progress has been made, a truly generative process planning solution has been elusive. This is due to the complexity of the problem involved. There is no uniform science base to support the research and development. Most of the studies are either fragmentary in terms of the approaches taken or *ad hoc* in nature. Frequently, the process planning task is not well defined and the coverage of the domain is incomplete. In addition to a proposed solution, this book provides a comprehensive technical survey of the important topics in computerized manufacturing process planning systems.

These topics include a historical review of the development of computerized manufacturing process planning systems, a fundamental consideration of manufacturing process planning, a discussion of the approaches of manufacturing process planning, including the definitions of manual and computerized process planning, and integration and intelligence issues. Implementation approaches are discussed in detail, ranging from the traditional variant group technology (GT) based process planning to the integrated and intelligent generative process planning. This book is virtually a one-volume encyclopedia on computerized manufacturing process planning systems: almost everything you always wanted to know about computer aided process planning (CAPP) and implementation.

The book starts with a very general view of what manufacturing industries, manufacturing processes, manufacturing systems, and manufacturing planning are. A broad variety of approaches to process planning from the traditional manual approach to current computerized approaches are examined, and the tasks of process planning are discussed. For any readers who may not be familiar with the traditional manual process planning, two typical examples are provided in Chapter 2 to facilitate their understanding of the fundamental requirements, such as the criteria, of process planning.

The ultimate goal of automated process planning is integrated computer aided design (CAD) and computer aided manufacturing

(CAM). The impact of computer integrated manufacturing (CIM) and the role automated process planning plays within CIM clearly deserve discussion in the book. Recent researches in the area of automated process planning have focused on the application of artificial intelligence (AI) techniques, such as expert systems (ES), knowledge-based systems (KBS), neural networks, and objective oriented programming techniques, all of which are thoroughly discussed in the book so that readers can understand how an intelligent process planning system performs. In the development of an integrated and intelligent process planning system for use in a real manufacturing environment, sophisticated technologies must be utilized in the development of automated process planning. Furthermore, automated tolerancing analysis is a critical issue for process planning, which has been overlooked by many researchers within the area. This book provides a comprehensive discussion of the automated tolerancing analysis from which readers can learn much.

One of the objectives of this book is to provide a state-of-the-art survey of the development of automated process planning, which involved an extensive discussion of eighteen well known families and/ or series of computerized manufacturing process planning systems, a large capacity collection of existing computerized manufacturing systems is provided in this book. This is the most comprehensive review and includes almost all of the existing process planning systems. These systems are developed by means of different approaches in different institutions and companies, some of them are academic research oriented, while some others are industrial practice oriented. They range from the laboratory development to those which are commercially available. The characteristics and general features of these systems are indicated in the book. The development institutions and time are also provided for those readers who may want to pursue further information about these systems. In order to learn how to implement a computerized process planning system, readers should not only have some fundamental knowledge about process planning and relevant issues of discussion about implementation, but they should also be able to examine a specific example and see how the development is done step by step. An example is provided in the last chapter of the book which forms a case study of implementation. The objective of the chapter is to provide an explicit outline of the development of a computerized manufacturing process planning system. Readers who are interested in developing their own process planning systems may follow the procedures set out in the chapter.

Anybody, from those interested in the research of automated process planning systems, to research and development (R&D) engineers from industrial companies, laboratories or institutions will find this book valuable. Of course the book is also very important for project directors, program managers, system supervisors, shopfloor foremen, division and department leaders, and even company presidents who are intending to supervise the integrated production.

Last, but not least, the book is written for college senior and graduate students in the following areas: manufacturing engineering, industrial engineering, mechanical engineering, system engineering, electrical engineering, and computer science areas. This book may be used for a one-semester course. In order to help instructors to review the contents of each chapter and enhance the learning results, a few practice exercises are provided in Appendix E. Instructors may select exercises according to their emphasis in the classroom lecture. I am also open to any feedback from any instructors who choose the book for a course in manufacturing process planning engineering.

The manuscript of the book has been used as a textbook for senior course IE 4352 – Manufacturing Engineering II in the fall of 1992 in the Department of Industrial Engineering, Texas Tech University. A part of the book has also been used in the graduate courses IE 5355 – Computer Aided Manufacturing, and IE 5352 – Advanced Manufacturing Engineering in the fall of 1992 and spring of 1993 respectively in the same Department. The feedback from students and instructors is very encouraging. Teaching results can be enhanced by providing a laboratory practice on the example XPLAN-R which is discussed in the book. If instructors are interested in laboratory practice, a c/c_{++} based demonstration model is available by contacting me at the Department of Industrial Engineering, Texas Tech University, Lubbock, Texas 79409-3061 USA. It is my hope that many engineering colleges and universities will be able to use the book as their textbook for some relevant courses.

Many friends, colleagues and students of mine contributed their efforts to this book. I would like to take this opportunity to express my deepest thanks for all of their selfless assistance and contributions. There are individuals who deserve special mention in this section for their help in providing reviews, ideas, suggestions, and/or proofreading the text. First, my thanks to Professor Leo Alting, my co-author and former adviser, for his brilliant encouragement, and endless support, as well as the many inspiring discussions which were major prerequisites for completion of this book. In fact the project was initiated by him, and without his efforts this book would never even have been planned. Secondly, my thanks go to Dr B. John Davies, professor emeritus from the University of Manchester Institute of Science and Technology, for his willingness to review the manuscript and provide invaluable suggestions. My special thanks go to Mr Jiannan Mei, one of my doctoral students in the Department of Industrial Engineering at Texas Tech University, who typed many chapters of the manuscript and drew almost all of the figures in the book. Without his significant contribution this book could never have been finished. Of course it would be unforgivable to forget to thank Dr Toben Lenue. The example XPLAN-R discussed in Chapter 11 is based on the early version of XPLAN which he developed in 1986 at the Institute of Manufacturing Engineering, Technical University of Denmark.

My special thanks go to my family, without whose tolerance during the difficult parts of the writing, the work would not have been accomplished. Especially, I want to thank my father Mr Gefei Xu for constantly encouraging me to write the book, and my wife Hongsi Xie for her support. Finally, I am most grateful to Mr Mark Hammond, my editor at Chapman & Hall, for his perseverance and patience with me during the manuscript preparation.

Hong-Chao Zhang
Lubbock, Texas
April, 1993

List of abbreviations

AI   Artificial Intelligence
AMPS   Automated Manufacturing Production Systems
AMS   Automated Manufacturing Systems
APPS   Automated Process Planning Systems
APT   Automatically Programming Tools

BOM   Bill Of Material

CAA   Computer Aided Analysis
CAD   Computer Aided Design
CAE   Computer Aided Engineering
CAG   Computer Aided Graphics
CAM   Computer Aided Manufacturing
CAP   Computer Aided Planning
CAPP   Computer Aided Process Planning
CAPP II   Computer Aided Production Planning
CAS   Computer Aided Scheduling
CIM   Computer Integrated Manufacturing
CNC   Computer Numerical Controlled
CP   Capacity Planning
CRP   Capacity Requirement Planning
CSG   Constructive Solid Geometry

DBMS   Data Base Management System
DNC   Direct Numerical Controlled
DMPS   Dedicated Manufacturing Production Systems
DSS   Decision Support System

EPPS   Expert Process Planning Systems
ES   Expert Systems

FBPF   Fuzzy Boundary Part Families
FEMA   Finite Element Method Analysis
FMC   Flexible Manufacturing Cell
FMPS   Flexible Manufacturing Production Systems
FMS   Flexible Manufacturing Systems

GPS	General Problem Solver
GT	Group Technology
IGES	Initial Graphics Exchange Specification
IIPPS	Intelligent and Integrated Process Planning Systems
IP	Integrated Production
IPP	Integrated Process Planning
JIT	Just In Time
KBES	Knowledge Based Expert Systems
LAN	Local Area Network
LTM	Logic Theory Machine
MD	Manufacturing Database
MPS	Master Production Scheduling
MRP I	Material Requirements Planning
MRP II	Manufacturing Resource Planning
NC	Numerical Controlled
PDES	Production Data Exchange using STEP
PP	Process Planning
QC	Quality Control
SDB	Single Data Base
STEP	Standard for The Exchange of Product data
TMS	Traditional Manufacturing Systems
TT	Type Technology
XPS	Experimental Planning System

Introduction to manufacturing (preliminary knowledge)

1.1 INTRODUCTION

In the last few decades, throughout the world, a large number of new technologies have appeared and have been implemented in the manufacturing industry. The new technologies include advances in microelectronics and computer technologies, new machine tools, robots, CVD (Chemical Vapor Deposition), PVD (Physical Vapor Deposition), plasma-aided manufacturing, etc. Among these technologies, the computer has had the most significant influence on all the functions of manufacturing enterprises. Computer Integrated Manufacturing (CIM) – a new manufacturing philosophy – has been proposed and addressed by many; for examples, see Merchant (1983, 1985, 1989), Alting (1986), Allen (1986), Bollinger (1988), Spur (1987, 1988a, 1988b), Yoshikawa (1988). CIM, the vision of the factory of the future, has been discussed for a long time. Integrated manufacturing systems are supposed to make up a major part of the factories of the twenty-first century.

In a narrow sense, a traditional production flow can be described as shown in Figure 1.1. It indicates a product creation through three main stages: the design stage, the process planning stage and the manufacturing stage. Utilizing the rapid developments in computer science, Computer Aided Engineering (CAE) techniques have been widely used in both design and manufacturing stages, i.e. Computer Aided Design (CAD) and Computer Aided Manufacturing (CAM).

In the CAD area many advanced tools have been developed. For instance, some Computer Aided Graphics (CAG) systems can provide many different geometrical models in terms of 2D or 2½ D wire-frames, surface descriptions and 3D solid modeling, etc. (Ryan, 1979). Furthermore, some CAG systems have been interfaced with Computer Aided Analysis (CAA) systems such as Mass Transfer and Finite Element Method Analysis (FEMA) systems. In this case, finite element meshes can be generated automatically and the data analyzed can be fed back

Figure 1.1 A traditional production flow of a manufacturing factory.

to the CAG system to optimize the model's geometrical shape (Krouse, 1982). In the CAM area, Numerically Controlled (NC) machines, Computer Numerical Control (CNC) machines and Direct Numerical Control (DNC) machines, as well as robotics, Programmable Logic Controller (PLC), inventory management systems, shop-floor data collection systems, and some Material Requirements Planning (MRP) systems have already been widely employed. However these isolated islands of automation need to be integrated to reach Integrated Production (IP). Although process planning plays a key role by linking design and manufacturing, it is still generally performed manually in industry. This is mainly due to the fact that traditional computer algorithms cannot perform process planning tasks that involve logical decisions that depend on production rules and expertise rather than on mathematical calculations. The development of Computer Aided Process Planning (CAPP) has been lagging behind that of CAD and CAM. CAPP did not receive much attention until about ten years ago.

Recently, much effort has been devoted to the development and implementation of Automated Process Planning (APP) systems. Many approaches have been explored, such as the variant and the generative approach of APP systems. Fairly high technologies, such as Artificial Intelligence (AI), Expert Systems (ES), Knowledge-Based Systems (KBS), Neural Networks, Object-Orientated Programming Techniques, etc., have been applied in the development of APP systems. Some new concepts of APP systems, such as Intelligent and Integrated systems, have been developed. This book addresses all these issues and systematically discusses them, beginning with fundamental theory and moving to some leading edge research activities. The aim of this book is to provide a comprehensive view of computerized manufacturing process planning from a manufacturing engineering perspective. The main emphasis is understanding the fundamental principles and implementation of APP systems. This book also provides explanations of the state-of-the-art computerized manufacturing process planning technologies. The criteria of how to evaluate an APP system are also addressed.

This book consists of 11 chapters, each of which addresses a specific issue of computerized manufacturing process planning.

Chapter 1, Introduction to Manufacturing, discusses some essential knowledge of manufacturing engineering. The manufacturing industry is identified, and the manufacturing system is defined. The functions of manufacturing processes and manufacturing planning are discussed.

Chapter 2, Manufacturing Process Planning, covers the different approaches to process planning, from conventional manual process planning to the current computerized manufacturing process planning technologies. This chapter places emphasis on procedures of manual process planning. Two manufactured parts, a rotational part and a prismatic part, are discussed as examples.

Chapter 3, Impact of Computer Integrated Manufacturing, covers most of the significant high-tech developments relevant to APP systems, such as Computer Aided Design (CAD), Computer Aided Manufacturing (CAM), Manufacturing Database Systems, Knowledge Based and Expert Systems.

Chapter 4, Group Technology in Process Planning, is dedicated to the discussion of Group Technology. GT is a backbone for computer aided process planning. It is so important that we need a whole chapter to discuss the fundamentals and practices of GT in CAPP.

Chapter 5, Automated Tolerance Analysis for Process Planning, addresses a new research area which is extremely important and which influences automated process planning systems. The fundamentals of the tolerance chain, tolerance chart, and analysis algorithms are discussed in terms of implementation in APP systems.

Chapter 6, Computer Aided Process Planning, provides the fundamentals of computer aided process planning theories. The new generation model of CAPP systems, the Integrated and Intelligent Process Planning (IIPP) system is discussed in this chapter.

Chapter 7, Implementation of CAPP Systems, discusses the state-of-the-art implementation of CAPP systems. Many different techniques for developing CAPP systems are discussed systematically. At the end of this chapter we discuss system architecture, which is a very important element in the development of CAPP systems. Some examples of existing architectures are illustrated at the end of this chapter.

Chapter 8, feature Based Techniques, is particularly focused on current research in the integration of design and process planning. Feature technologies play an important role in the integration of design and process planning. We are going to introduce a few feature recognition methods in this chapter.

Chapter 9, A Survey of CAPP Systems, describes the state-of-the-art development of CAPP systems. Eighteen well-known CAPP systems are discussed, including their advantages and disadvantages. In the last section of this chapter, evaluation criteria for CAPP systems are discussed.

Chapter 10, Crucial Issues and the Future Trends of CAPP Development, presents some issues which may influence the future trends in the development of CAPP systems. These issues include software, hardware, user friendliness, interfacing, integration ability, and some other relevant topics.

Chapter 11, Building a CAPP System – A Case Study, uses an actual case to demonstrate how to build a CAPP system. This specific system is for rotational parts, but the basis of this development can be extended to explain the development of a CAPP system for other kinds of geometrical shape families.

This book can be used as a textbook for senior undergraduate or graduate students in departments of mechanical engineering, manufacturing engineering, manufacturing technology, or industrial engineering for their manufacturing process planning course or for an advanced manufacturing engineering course. This book can also be used as a reference book for guiding research activities in the area of the development of CAPP systems and production integration. Project managers, production managers, manufacturing engineers, mechanical engineers, and industrial engineers may also be interested in this book as a means to update their knowledge and facilitate their routine jobs.

1.2   MANUFACTURING INDUSTRIES

What is 'industry'? According to Webster's Dictionary, industry means 'Any department or branch of art, occupation, or business; especially one which employs much labor and capital and is a distinct branch of trade'. This explanation may cover a very wide variety of enterprises as industry. However, the interests of engineers may focus on the enterprises whose products can be traded on the major markets, namely the production industry.

In general, production industries can be divided into two main categories, depending on the nature of their production operations. They are the manufacturing industries and the process industries. Manufacturing industries are typically identified with discrete item production, such as cars, computers, machine tools, machines used for various purposes, ships, weapons, aircraft, appliances and the components that go into these products. The process industries are represented by various chemicals, plastics, petroleum products, food processing, soaps, steel, paper making, and cement. The focus in this book is on the discrete part manufacturing industry.

There is an important factor which characterizes the nature of the production activity according to the quantity of product made, that is, the production type. In general there are three types of production:

1. Job shop production
2. Batch production
3. Mass production

Table 1.1 Approximate quantities for different production types

Size of product		Large size product	Middle size product	Small size product
Job shop production		fewer than 5	fewer than 20	fewer than 100
Batch production	Small	5–100	20–200	100–500
	Middle	–	200–500	500–5000
	Large	–	500–5000	5000–50000
Mass production		–	over 5000	over 50000

The approximate quantities for each type of production are illustrated in Table 1.1.

Job shop production is low volume. The manufacturing lot sizes are small; usually fewer than ten. This production is commonly used to meet specific customer orders, and there is a great variety in the type of work the company must do. The production equipment for job shop production is flexible and general-purpose. The most up-to-date equipment for this type of production is programmable. This type of production usually requires relatively highly skilled workers to perform a range of different work assignments. Examples of products manufactured in a job shop include space stations, aircraft, machine tools, special tools and equipment, as well as prototypes of future products.

Batch production involves the manufacture of medium-sized lots of the same item or product. The lots may be produced only once or at regular intervals. Batch production is used mainly to meet the continuous demand for one product from one or several regular customer(s). Thus, the shop floor produces in order to build up an inventory of the item, then it changes over to other orders. When the stock of the first item becomes depleted, production is repeated to build up the inventory again. The manufacturing equipment used in batch production is general-purpose but designed for higher rates of production. The machine tools used in batch production are often combined with specially designed jigs and fixtures that increase the production rate. Recently, considerable Numerical Control (NC), Computer Numerical Control (CNC), and Direct Numerical Control (DNC) techniques have been applied in this type of production. The most up-to-date equipment for batch production is programmable and flexible. Cellular Manufacturing System (CMS) is a typical production model for batch production. Examples of items made in batch-type production include industrial equipment, furniture, hand tools, kitchen wares, some high quality consumer appliances, and component parts for many assembled consumer products. With the pressure of competition in the world market and application of new technologies, many companies have changed their traditional unique mass production into multiple variety and lot batch size production. For example most automobile companies produce many different models for their annual production.

It has been estimated that perhaps as much as 75% of all parts manufacturing is in lot sizes of 50 pieces or fewer. Therefore, batch production and job shop production constitute an important portion of the total manufacturing activity.

Three categories of batch production can be distinguished:

1. Small batch
2. Middle batch
3. Large batch

Table 1.1 shows the corresponding batch production quantities.

Mass production is the continuous specialized manufacture of identical products. Mass production is characterized by very high production rates, equipment that is completely dedicated to the manufacture of a particular product, and very high demand rates for the product. Not only is the equipment dedicated to one product, but the entire plant is often designed for the exclusive purpose of producing that particular product. The equipment is special-purpose rather than general-purpose. The investment in machines and specialized tooling is high. In some sense, the production skill has been transferred from the operator to the machine. Consequently, the skill level of labor in a mass production plant is usually lower than in a batch plant or job shop.

Two categories of mass production can be distinguished:

1. Quantity production
2. Flow type production

Quantity production involves the mass production of single parts on fairly standard machine tools such as punch presses, injection molding machines, and automatic screw machines. These standard machines have been adapted to the production of the particular part by means of special tool-die sets, molds, and form cutting tools, respectively, each designed for the part in question. The production equipment is devoted full time to satisfying a very large demand rate for the item. In mass production, the demand rate and the production rate are approximately equal. Examples of items in quantity production include components for assembled products that have high demand rates (automobiles, some household appliances, light bulbs, etc.), hardware items (such as screws, nuts, and nails), and many plastic molded products.

Table 1.2 Characteristics of different production types

Characteristics of products	Job shop production	Batch size production	Mass production
Production objects	changed frequently	changed widely	relatively fixed
Equipment and layout	general purpose and process oriented	general equipment with special fixtures, flow type oriented	dedicated equipment with dedicated fixtures, automated fixed flow line
Fixtures	general purpose fixtures, usually in company with machine tools	normally special fixtures	automated dedicated fixtures
Tooling and measuring tools	general oriented	special oriented	dedicated oriented with high efficiency
Workpiece mutual exchange	try and modify	normally using mutual exchange	mutual exchange only
Stock	wood model casting and free forging	metal model casting and die forging	metal model and machinery casting, and die forging with highly efficient machines
Labour skill	higher	middle	lower
Production rate	lower	middle	higher
Cost/piece	higher	middle	lower
MLT/piece	longer	middle	shorter

Flow type production is the other category of mass production. The term suggests the physical flow of the product in oil refineries, continuous chemical process plants, and food processing. While these are examples of flow production, the term also applies to the manufacture of either complex single parts (such as automotive engine blocks) or assembled products. In these cases, the items are made to 'flow' through a sequence of operations by material handling devices (conveyors, moving belts, transfer devices, etc.). Examples of flow production include automated transfer machines for the production of complex discrete parts and manual assembly lines for the assembly of complex products.

Table 1.2 illustrates some important characteristics of different production types, broken down by production quantity. As a matter of fact, the production ranges of the three major categories overlap to some degree, simply because it is difficult to figure out a clear dividing line between the different types.

1.3   MANUFACTURING SYSTEMS

The term 'manufacturing systems' was probably first proposed by Dr M. Eugene Merchant in the 1960s. At that time, with computers emerging in industry, many new objectives were evolving based on the application of computers in industry such as 'unification and co-ordination of research aimed at the scientific-engineering approach to production', as well as 'automation of the entire manufacturing process from initial design to finished product'. In order to achieve these objectives, visualization and study of the manufacturing process as a whole are required. Merchant stated, 'this, we propose, is best accomplished by applying the systems concept to manufacturing and using this concept, then, as the broad program guide for production engineering research. Thus, the concept of the 'manufacturing system' is already beginning to be used by the more advanced production engineering research establishments in the United States as the overall framework for planning and coordinating their research activities' (Merchant, 1962). Since then, in the last three decades, the term manufacturing systems has been further defined and many new concepts have evolved.

Manufacturing systems can be defined as systems that perform a sequence of transformation processes to convert the initial ideas of product design into realistic finished products that have value in utilization and in the market-place. The transformation processes, which are performed by a combination of manual labor, machine tools, and energy can be controlled manually or automatically. Such a manufacturing system may be described as shown in Figure 1.2.

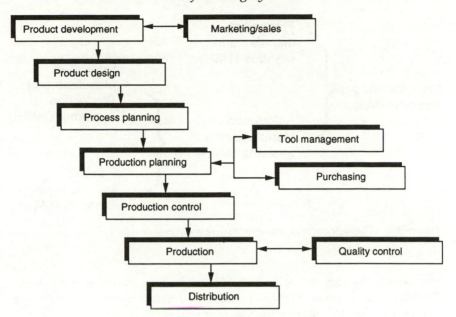

Figure 1.2   Manufacturing system diagram (Lenau, 1989).

Manufacturing systems are usually controlled either manually or automatically. According to which method is used to control the manufacturing systems, the manufacturing systems can be classified as either Traditional Manufacturing Systems (TMS) or Automated Manufacturing Systems (AMS), respectively.

The primary characteristic of TMS is a comprehensive application of manually assisted equipment. Manual work often requires large batches of workpieces to justify starting up a process and operation, since the setup of equipment takes a long time in comparison to the process times for each workpiece (Larsen, 1989).

Automated Manufacturing Systems (AMS), also called Automated Manufacturing Production Systems (AMPS), consist of an integrated assembly of machines and/or associated equipment necessary to carry out production with a minimum amount of manual intervention, together with the means for transferring components automatically through the system, and all operating under fully programmable control. An automated manufacturing system usually includes machine tools for machining or forming materials, but other machines, such as those for cleaning, assembly and painting of the product, are also included. To run the system, Computer-Aided Scheduling (CAS), Computer-Aided Planning (CAP), and other computer based tools are applied.

Two main types of automated manufacturing production systems

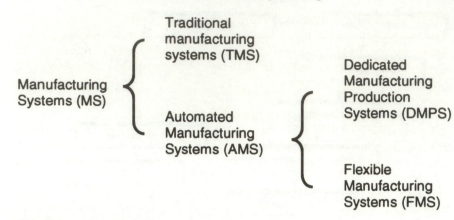

Figure 1.3 Classification of manufacturing systems.

can be distinguished: Dedicated Manufacturing Production Systems (DMPS) and Flexible Manufacturing Production Systems (FMPS) or simply Flexible Manufacturing Systems (FMS).

Dedicated Manufacturing Production Systems (DMPS)

A dedicated manufacturing production system is an automated system designed for the production of one product only, and which cannot readily be adapted for the production of other products.

Flexible Manufacturing Systems (FMS)

A flexible manufacturing system is an automated system which is capable of producing any of a range or family of products, with a minimum amount of manual intervention. The flexibility is usually restricted to the family of products for which the system was designed. Figure 1.3 illustrates the classification of manufacturing systems.

1.4 MANUFACTURING PROCESSES

The term 'process' is defined by Professor L. Alting as 'a change in the properties of an object, including geometry, hardness, state, information content (form data), and so on. To produce any change in property, three essential agents must be available: (1) material, (2) energy, and (3) information. Depending on the main purpose of the process, it is either a material process, an energy process, or an information process.' (Alting, 1982). Furthermore, material processes can be classified into three categories as follows:

Mass-conserving processes correspond to through flow.
The mass of the initial work material is equal to (or nearly equal to) the mass of the final work material, which means, when referring to geometrical changes, that the material is manipulated to change its shape.

Mass-reducing processes correspond to diverging flow.
The geometry of the final component can be circumscribed by the initial material geometry, which means that a shape change is brought about by the removal of material.

Assembly or joining processes correspond to converging flow.
The final geometry is obtained by assembling or joining components so that the mass of the final geometry is approximately equal to the sum of the masses of the components which are manufactured by one or both of the previous methods.

Table 1.3 illustrates the further classification of manufacturing processes relative to these three categories.

Table 1.3 Classification of manufacturing processes (only typical process examples are mentioned) (Alting, 1982)

Process or flow type	State of material	Category of basic process	Primary basic process	Process examples
Mass-conserving processes $(dM = 0)$	solid	mechanical	plastic deformation flow and	forging and rolling
	granular	mechanical	plastic deformation	powder compaction
	fluid	mechanical	flow	casting
Mass-reducing processes $(dM < 0)$	solid	mechanical		turning, milling, and drilling
		thermal	melting and evaporation	electrical discharge machining (EDM) and cutting
		chemical	dissolution	electro-chemical machining (ECM)
			combustion	cutting
Joining processes				
Atomic bonding	solid	mechanical	plastic deformation	friction welding
	fluid (vicinity of the joint)	mechanical	flow	welding (fusion)
Adhesion	solid (fluid filler material)	mechanical	flow	brazing

According to the sequences of manufacturing flow, manufacturing processes can also be classified into one of the following four categories:

Initial forming processes

Initial forming processes create the initial forms of the work material, for example casting and plastic molding. In these cases, the raw materials are converted into the basic geometry of the desired product. Additional processing is usually required to achieve the final shape and size of the workpiece.

Consecutive processes

Consecutive processes usually follow the initial forming processes and create the final desired geometry of the workpiece, for example machining (turning, milling, drilling, grinding etc.) and press-working (blanking, forming, drawing etc.).

Physical property enhancement processes

Processes which enhance physical properties do not significantly change the physical geometrical shape of the workpiece. Instead, the physical properties of the material are improved in some way. Heat treatment, shot peening and cold-work hardening are some examples in this category. Note that processes which enhance physical properties do not always follow consecutive processes. Sometimes they may be performed between two consecutive processes. For example, a heat treatment process (hardening), if desired, is always done prior to a grinding process.

Completive processes

Completive processes are usually the final processes required for a workpiece or a finished product. The purposes of completive processes are to improve the appearance, or to provide a protective coating on the part, or even to assemble a final product. Examples in this category include polishing, painting, chrome plating, and assembling. Figure 1.4 illustrates this general process model.

1.5 MANUFACTURING PLANNING

Planning, in a generic sense, is an instinctive part of intelligent behavior, often performed subconsciously by human beings. Planning can be viewed as the activity of devising means to achieve desired

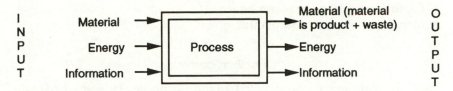

Figure 1.4 A general process model.

goals under given constraints and with limited resources. Thus, three basic components of any planning activity are: goals, constraints, and resources (Ham and Lu, 1988).

Manufacturing planning can be seen as the coordinating function that links the information transformation from the stages where the objectives are generated to the stages where the objectives are realized within a manufacturing system. In any manufacturing environment, specific planning functions take place at different stages, and the global sum of these activities constitutes a complete manufacturing planning system. There are many factors (e.g., technical, economic, social etc.) that must be taken into consideration in achieving an effective overall manufacturing plan.

In general, manufacturing planning activities can be classified into five categories as follows.

Production planning

In many cases it is often called master production planning, production scheduling, or master production scheduling, or even master scheduling; by any name it is a listing of the products to be made, when the products are to be delivered, and in what quantities. Units of time (usually months) are often used to specify the deliveries on the master production plan.

Requirement planning

Requirement planning is sometimes also called activity planning or a balance sheet. The objective of requirement planning is to plan manufacturing activities accurately by calculating the net requirement in conjunction with production planning (the master schedule). Some lower level activities within requirement planning are:

- Individual components and subassembly requirement planning
- Purchase parts requirement planning
- Materials requirement planning (MRP)

- Facilities requirement planning
- Manpower requirement planning

Capacity planning

Capacity planning is the detailed planning phase. It is a scheduling and sequencing task. The major objectives of capacity planning are:

- Meeting delivery dates
- Keeping to a minimum the capital tied up in production
- Reducing manufacturing lead time
- Minimizing idle times (machine out of work) on the available resources
- Providing management with up-to-date information and solutions

Some of these objectives conflict with each other, and some relate to the activities of requirement planning.

Process planning

The purpose of process planning is to select and define, in detail, the processes that have to be performed in order to transform the raw materials into a given shape. The primary objective is to define feasible processes. Cost and throughput are secondary objectives, and available resources (machine tools, cutting tools and labor) act as constraints. The activities of process planning will be discussed in detail in the next chapter.

Operation planning

In some of the literature, operation planning, as a lower level planning activity of process planning, has also been classified into the manufacturing planning categories (Ham and Lu, 1988). Operation planning, often called operation sequencing, is used to determine the details of the parameters that will ensure the smooth completion of planned manufacturing operations. It takes into account the specific cutting sequences, cutting steps, cutting depths, feed rates, etc., for each specific workpiece that is operated, on each specific machine, with each specific cutting tool. The process tolerances are considered within the operation planning. In this book the operation planning, named operation sequencing, will be included in the activities of process planning.

REFERENCES

Allen, D.K. (1986) Architecture for Computer-Integrated Manufacturing. *Annals of the CIRP*, **35**(1).

Allen, D.K. (1987) An introduction to Computer-Aided Process Planning. *CIM Review*, Fall 1987.

Alting, L. (1982) *Manufacturing Engineering Processes*, Marcel Dekker Inc., New York, p. 367.

Alting, L. (1986) Integration of engineering functions/disciplines in CIM. *Annals of the CIRP*, **35**(1).

Alting, L. and Zhang, H. (1989) Computer Aided Process Planning: The state-of-the-art survey. *International Journal of Production Research*, **27**(4), pp. 553–85.

Bollinger, J.G. (1988) The factory of the future: Technology aspects. *Annals of the CIRP*, **37**(2).

Ham, I. and Lu, S.C.-Y. (1988) Computer-Aided Process Planning. The present and the future. *Annals of the CIRP*, **37**(2).

Krouse, J.K. (1982) *Computer-Aided Design and Computer-Aided Manufacturing*, Marcel Dekker Inc., New York, p. 147.

Larsen, N.E. (1989) *Simulation: A concept for production scheduling*. Ph.D. thesis, Institute of Manufacturing Engineering, Technical University of Denmark. Publication No. PI.89.01-A/AP.89.01, p. 398.

Lenau, T. (1989) *Knowledge based engineering design and process planning*. Ph.D. thesis, Institute of Manufacturing Engineering, Technical University of Denmark. Publication No. PI.89.08-A/AP.89.07, p. 175.

Merchant, M.E. (1962) The manufacturing-system concept in production engineering research. *CIRP-Annals*, **X**, 1961/1962.

Merchant, M.E. (1983) *The Factory of the Future*. Reported in Technical University of Denmark, 1983, Lyngby.

Merchant, M.E. (1985) Computer-Integrated Manufacturing as the basis for the factory of the future. *Robotics and Computer-Integrated Manufacturing*, **2**(2), pp. 89–99.

Merchant, M.E. (1989) Important Trends in the International Development of Manufacturing Systems Science and Technology. Proceedings of 21st CIRP seminar on Manufacturing Systems – Tool for Man, June 1989, Stockholm.

Ryan, D.L. (1979) *Computer-Aided Graphics and Design*, Marcel Dekker Inc., New York, p. 324.

Spur, G. and Specht D. (1987) Computer Integrated Manufacturing in future factories. *Robotics and Computer-Integrated Manufacturing*, **3**(2), pp. 147–55.

Spur, G. (1988a) The factory of the future: Management aspects. *Annals of the CIRP*, **37**(2).

Spur, G. and Mertins K. (1988b) State of the Art and Trends Toward Computer Integrated Manufacturing. Proceedings of the 7th PROLAMAT conference, Dresden, GDR, June 14–17, **II**.

Yoshikawa, H. (1988) The factory of the future: Social aspects. *Annals of the CIRP*, **37**(2).

Zhang, H.C. (1987) *Overview of CAPP and development of XPLAN*. Institute of Manufacturing Engineering, Technical University of Denmark. Publication No. AP.87-26.

CHAPTER 2

Manufacturing process planning

2.1 INTRODUCTION

Process planning is performed in virtually all industries: its significance is greatest in small-batch, discrete part, metal fabrication manufacturing industries. Recently, however process planning has also been recognized as playing an important role in other manufacturing and process industries, such as electronics manufacturing companies, furniture manufacturing companies, or even chemical process plants. Process planning is the transformation of part design specifications from detailed engineering drawings into operating instructions for necessary manufacturing.

A completed manufacturing process includes the whole transformation from a raw material stock into a desired product part or component. In general, manufacturing process planning refers to either machining process planning or assembly process planning. Machining process planning is concerned with how each single workpiece is machined on individual machines or manufacturing cells, while assembly process planning is concerned with how several workpieces can be assembled together to form a machine part. Machining process planning is often simplified as process planning. In this book, the term process planning is used only to represent machining process planning, and the word 'assembly' will be always used as a prefix to distinguish assembly process planning. Several terms which are often used in process planning need to be defined before further discussion.

Process: Process is the basic unit for constructing process plans. A process can be defined as a procedure in which one or one group of workpiece(s) is continually machined on one machine or workstation by one or one group of operator(s), for example a rough turning process on a lathe, a finish turning process on a lathe, a milling keyway process on a milling center, or a drilling hole process on a drilling machine.

Operation: Operation is the sub-unit of a process, and it is the basic part of the process. Each operation is accomplished without changing cutting conditions (speed, depth, and feed), cutting tools, or cutting surfaces. Usually, one process consists of several operations. These operations are sequenced in a desired order called operation sequencing, which is an integral part of process planning.

Cut: Cut is the sub-unit of an operation, and it is the basic part of the operation. A cut can be defined as a procedure in which the cutter passes the cutting surface once. In other words, some operations need several cuts, with the same cutting tool and under the same cutting conditions (speed and feed). In this case, each pass of the cutting surface by the cutting tool is called a cut. Usually one operation consists of one or more cuts, depending on the required cutting volume.

2.2 DEFINITION OF PROCESS PLANNING

Process planning can be defined as the systematic determination of the detailed methods by which workpieces or parts (if considering an assembly) can be manufactured economically and competitively from initial stages (raw material form) to finished stages (desired form). Geometrical features, dimensional sizes, tolerances, materials, and finishes are analyzed and evaluated to determine an appropriate sequence of processing operations, which are based on specific, available machinery or workstations. In general, the inputs to process planning are design data, raw material data, facilities data (machining data, tooling data, fixture data, etc.), quality requirement data, and production type data. The output of process planning is a process plan. The process plan is often documented into a specific format and called a process plan sheet. Process plan sheets may be referred to by different names, such as process sheets, operation sheets, planning sheets, route sheets, route plans or part programs. A process plan is an important document for production management. The process plan can be used for the management of production, the assurance of product quality, and the optimization of production sequencing. The process plan can even be used to determine equipment layout on the shop floor; recent research results have also demonstrated that process planning plays an important role in a flexible manufacturing system (FMS) and computer integrated manufacturing enterprises. Process planning is the key link for integrating design and manufacturing. Because a process plan is such an important document, everyone must respect and execute it seriously. In developing a new product, the process plan provides necessary information for technical and equipment preparation, such as tools, jigs and fixtures, machines, inspection devices, raw material stocks, inventory plans, purchasing plans, personnel requirements, etc. In designing a new factory or extending or modifying an old factory, the process plan is essential information that will determine equipment requirements, area of shop floor occupation, and investment.

The following essential information is necessary for process planning.

Design data, which includes all the assembly and separated single part drawings. Usually, design data is presented in the form of blueprints. However, if the design work is done on a computer, then the design data is presented in the form of CAD models.

Quality requirement data, which will affect the tools, fixtures, and equipment selection of process planning.

Production type data, which may lead to different process plans for the same product with different production types. While mass production requires a strategy of process distribution, job shop and batch size production prefer a strategy of process concentration.

Raw material data, including information from raw material storage and capacity to the capability of making adequate stock and blank of the company.

Company capacity and capability data, such as equipment, tools, fixtures, dedicated machines, general machines, machine calls, stations, and machining centers, as well as FMS.

Figure 2.1 illustrates a process planning model, which indicates the input to process planning and output from process planning.

2.3   APPROACHES TO PROCESS PLANNING

In general, approaches to process planning can be classified into two main categories: manual process planning and computerized process

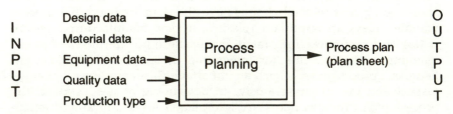

Figure 2.1   A process planning model.

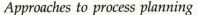

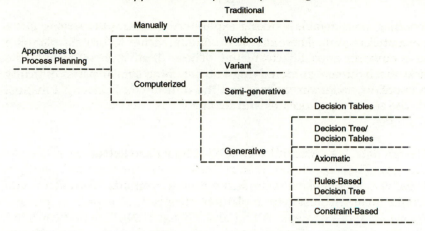

Figure 2.2 Classification of approaches to process planning.

planning. Figure 2.2 illustrates a hierarchical classification tree of approaches to process planning. The approaches to computer aided process planning will be discussed in Chapter 4. The approaches to manual process planning include two distinctive categories, the traditional approach and the workbook approach (Allen, 1987).

Traditional approach

The traditional approach to process planning involves examining the information of a part design described in the form of a blueprint, identifying similar parts (from memory or from a code book), and manually retrieving process plans for these similar parts. A new process plan is then created by modifying and adapting the old one to meet the special requirements of the new print. Customarily, the process planner will consult the foreman in the production shop to find out how the part is actually being processed.

Workbook approach

An alternative and more efficient approach to process planning is constructing a workbook containing a menu of pre-stored sequences for operations for given types of workpieces. These stored process groups may be quickly selected and sequenced by the process planner. The menu selections are then typed on the regular process sheet and reproduced as required.

The main advantages of manual approaches are low investment and flexibility. For the workbook approach, a well-trained planner can produce a large number of process plans for simple parts. For both approaches, considerable working experience is crucial to carrying out a good process plan. Therefore, manual approaches to process planning have some obvious disadvantages. These disadvantages include the lack of consistency in identifying and planning similar parts, difficulty in specifying common tooling, and the difficulty of updating a manual file to reflect new processes and tooling.

2.4 MISSION OF PROCESS PLANNING

Usually, the person who produces process plans is referred to as a process planner or simply a planner. The task of a process planner involves a series of steps (Alting and Zhang, 1989). The first consideration is the interpretation of the design data, which is usually displayed either in traditional blueprints or more recently by CAD models from computer aided design systems. In this stage two major tasks need to be accomplished: (1) understanding the functions, conditions, and specifications of the product, clarifying the relative assembly positions and mutual functions, and estimating the suitability of the design requirements; (2) examining and analyzing the design data by carefully reading through the assembly and parts drawings. In this step, required information, such as production types, geometrical configuration, raw material properties, dimension tolerances, surface roughness, heat treatment and hardness, as well as some special requirements, will be studied and interpreted. It is important to check out if the design is completed, the design requirements are reasonable, part configurations are manufacturable, all dimensions and tolerances are available, and the surface roughness and tolerances are appropriate; it is also important to find out if the design is optimal from the manufacturing point of view. If problems are detected at this stage, the planner must discuss the problems with designers so that both sides are in agreement on how to modify or remedy the design. In traditional manufacturing companies, this work may sometimes cause disagreement between designers and planners because of conflicting strategies and policies for design and manufacturing. In order to avoid the hassles, many manufacturing enterprises have insisted on removing the barrier between design and manufacturing (Tuttle, 1983). This is a topic of recent research activities, namely concurrent or simultaneous engineering. Of course, the potential benefit of applying concurrent engineering is far greater than merely solving conflicts between design and manufacturing. Concurrent or simultaneous engineering is outside the scope of this book; however, interested readers may refer to other relevant publications.

The second step of process planning is to design stock. Usually, the properties of the raw materials are specified by the design engineer, because these properties are dependent upon the requirement of a product's design. However, the geometrical shapes of raw materials, namely stock or, in other terms, blanks, are designed by process planners. The stock design is usually based on the final geometrical shape of the part and the production types. Some selection of properties of materials may affect the stock design or even the selection of processes. For instance, if the raw material of nodular cast iron is selected by the designer for a specific gear part, then a casting process must be selected for stock design. In this situation, the planner needs to discuss the stock design with the foundry shop. However, if medium-carbon steel is selected for the same gear part, the stock design may be quite different from the first design. The criteria of stock design are (1) assuring the quality of stock, (2) minimizing machining requirements, (3) increasing the utilization of material, and (4) reducing the manufacturing cost and lead time.

According to the product design and stock design, the data of machining processes, such as turning, milling, drilling, grinding, etc., that transform the raw material into the desired part, must be selected. In the meantime, the datum surfaces for fixturing must be determined, which is the key step in implementing an optimal process plan. Sometimes, it is necessary to implement several different plans for the same parts. Comparison must be made according to the specific condition and capacity of the manufacturing environment. Once the processes are selected, the machining equipment (machine tools, workstations, machining centers, etc.) which can carry out one or more machining processes, have to be selected. The selection should take into consideration availability, process capability (size, accuracy, etc.), range of machining operations, production rate, etc. In order to complete the entire optimal production, the process planner must consider the information from the production scheduling and sequencing for the machine tool selection. For modern FMS and integrated manufacturing environments, the selection of machine tools has become even more important. Some recent research activities for integrated process planning with production scheduling are being conducted for this purpose (details are given in section 10.6.3).

After the machine tools are selected, the cutting tools, including clamping and measuring devices and auxiliary tools, must be chosen. If some dedicated tools, fixtures, or auxiliary devices need to be made, the design project must be proposed by the process planner.

After all the tools have been selected, then the operation sequence must be determined. The determination of an operation sequence is usually based on a company-specific strategy which consists of comprehensive operations for a defined group of parts. Each operation is described by selection criteria which depend on the shape and dimensions of the part.

In the operation sequencing step, the amount of material to be removed by each process and operation must be checked. While finish machining requires only a small amount of removal of materials, rough machining requires relatively more. Usually it is necessary to calculate the overall amount of material removal for operations. The most efficient method to determine the amount of material for removal is operational dimensioning and tolerance chain analysis that will provide the precise operation dimensions and tolerances of removals. The automated operational tolerance chain theory will be discussed in Chapter 4. The appropriate cutting conditions, such as depth of cut, feed, and speed rates, then have to be determined and the total machining and non-machining times have to be calculated, including batch setup time, loading and unloading, tool changing, and inspection time. The cost of the process can be also included, if desired. Finally, process plans are made and edited. Checking for syntax and path errors should be done in this stage. Significant economic benefits can be obtained if optimum production concepts are adopted throughout the process planning.

In summary, the tasks of process planning can be itemized as follows:

- Interpretation of product design data
- Stock design
- Selection of machining processes
- Selection of machine tools
- Determination of fixtures, tools, and datums
- Sequencing of operations
- Determination of operational dimensions and tolerances
- Determination of the proper cutting conditions
- Calculation of the total times
- Generation of process sheets, including NC data

2.5 MAKING A PROCESS PLAN

In order to implement an optimal process plan, the process planner must consider many factors which are involved in manufacturing; among all these factors, datum selection is very important, especially named setup datums. The selection of datums in machining is closely related to the specification of the operation sequence. When a process plan is to be made, the relationships between positioning and the sequence and its influence on machining accuracy have to be carefully considered. In the following sections, we are going to discuss the concept of datums.

2.5.1 Datums

A part consists of several surfaces which have specified mutual (positional and dimensional) relationships to guarantee proper functions.

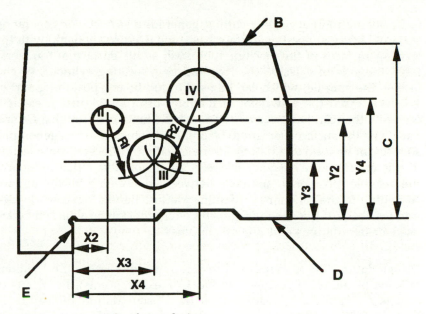

Figure 2.3 An example of part design.

Therefore, in the machining process, one or more surfaces on the part should be used as datum elements to machine other surfaces and satisfy the requirements on the blueprint. The concept of the datum is introduced to describe this mutual positional relationship. In order to ensure these relationships, positional tolerance is often applied. In Figure 2.3 the relative positioning between feature D (plane) and feature IV (hole) is indicated. To assure this positional requirement during the machining process, an appropriate setup datum must be selected. First it is necessary to define datums. Datums can be defined as the features (surfaces, lines, points) which can be used to coordinate other features (surfaces, lines, points). Feature D in Figure 2.3 is a datum. When the positions of these features are to be calculated or measured, the datum is the starting point of the calculations or measurements. The concept of datum is meaningful only where the relative positional relationships among the points, lines, and surfaces are concerned.

The relative positional relationship with the datum includes parallelism, perpendicularity, and concentricity (coaxiality). Positional tolerances may be specified on the blueprint if the functional or technological requirements are high, otherwise the relationship will be restricted by the dimensional tolerance. Different machine tools can make different surface elements (features) on a part. These features can be regarded as the loci of the tool on the part. By cutting within the machining zone on the machine tools, all the achievable loci (programmable on an

NC machine tool) are called cutting generating surfaces or generating surfaces. Every generating surface has some fixed relationship with the locating surfaces of the machine tool, such as the surface of the work-table, the side of a T-slot, the locating cone, locating cylinder, or end face on the spindle, which can be used to locate and position a part or a fixture. From the viewpoint of manufacturing, any surface element on a part that can have a fixed relationship (direct or through a fixture) and make the surface element to be obtained coincide with a generating surface can be used as a datum. In the machining process, the selection of a datum from the elements is one of the major factors that influences the relative positional accuracy between the elements on a part. According to the functions of datums, datum elements can be divided into two basic types according to the environments in which they are used: design datums and process datums.

Design datums are features used to determine and assure the positions of some of the other features (surfaces, lines, or points), by means of design. In Figure 2.3, feature D is the design datum of feature B. Feature IV (hole) has two design datums, feature D and E, respectively, in the vertical and horizontal directions. The design datum of feature III (hole) is made up of the axial center lines of features II (hole) and IV (hole) and the dimensions R1 and R2. In general, a design datum is used on the blueprint to specify the dimensions or relative positions of other elements, such as parallelism, perpendicularity, or concentricity. It is assigned by the designer based generally on the function of the part in the product and more specifically on the feasibility of machining. For an individual part, there may be quite a few specifications for the elements, but usually there is only one major design datum in each direction, where this design datum is usually the datum that is used to locate the part in the assembly of the product.

Process datums are features (surfaces, lines, points) which are used to clamp the workpiece for machining other features. According to the different functions, process datums can be further classified into three categories: setup datum, measuring datum, and assembly datum.

Setup datums (also referred to as positioning datums) are features (surfaces, lines, and points), which form definite positions between operational dimensions and cutting tools while the workpiece is clamped on a machine or a fixture. For instance, Figure 2.3 indicates that feature D is the setup datum for machining feature IV (hole). Consequently, feature D is also the design datum of feature IV (hole). In this case, the setup datum coincides with the design datum. Datum coincidence will increase manufacturing accuracy, reduce manufactur-

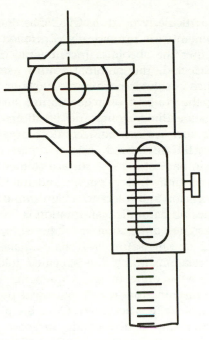

Figure 2.4   An example of a measuring datum.

ing errors, and simplify operation sequencing without changing any other manufacturing factors. As a matter of fact, datum (design and setup) coincidence is a desired situation for optimal accuracy in process planning.

Measuring datums are features (surfaces, lines, points) which are used for measuring operational dimensions. Figure 2.4 illustrates the measuring datum. For a process or an operation, the measuring datum should coincide with the positioning datum in order to check that the parts are made within the operational specifications. If the positioning datum does not coincide with the design datum, the measurement may give false results.

Assembly datums are features (surfaces, lines, points) which are used for the determination of the assembly position of each workpiece within the entirety of the product. Features D, E, IV, III, and II in Figure 2.3 are assembly datums.

2.5.2   Synchronous datums

Besides the datums described in section 2.5.1, there is another datum named the synchronous datum. Actually, synchronous datum belongs to setup datum but is particularly used on CNC machines. On an NC machine, the tool movements are numerically controlled through programs. No matter whether the absolute, incremental, or the floating coordinate system is selected, the nominal values used in the NC program are converted to absolute coordinate values in the controller of the machine. Using the concept of a generating surface, one can clearly see that the relative dimensional and positional relationships between the generating surfaces can be accurately programmed. The relationship can be exactly duplicated onto the part if the surface elements are machined in one setup. The surface elements obtained in one setup are mutually datumed, dependent, and interrelated; therefore, the concept of dimension and tolerance chain cannot be used here and is not necessary since no datum transformation is introduced. That is to say that for any designed dimension or relationship, if the surface elements are machined in one setup, no chain analysis is needed for that dimension or relationship. If the specified tolerance cannot be obtained, nothing can be done in the sequencing unless a new machining method or a machine tool with a narrower process capacity is adopted or a high rate of scrap is accepted. Once the part is removed and relocated, this relationship is destroyed. (In some special cases, where a part is center-supported and the repeatability of the setup is extremely accurate, the setups can be regarded as one.) These mutually datumed elements are called synchronous datum elements. Other datum elements obtained in different setups are also called synchronous datum elements, where some errors such as fixture error, tool calibration error, tool wear, and setup error are introduced into the

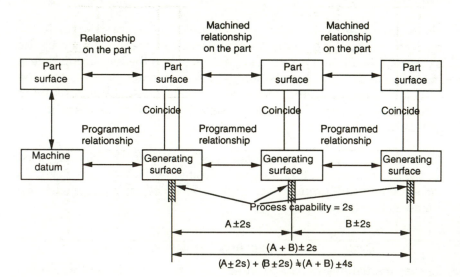

Figure 2.5   Relationships obtained in one setup.

part dimensions. This concept has not been addressed by any researchers but is depicted in Figure 2.5.

2.5.3 The relationship between the datum elements

Sometimes the design datum cannot be used both as a measuring datum and as a positioning datum. In other words, it may be difficult to make these datum elements coincide with each other. Therefore, it is very important to make their relationship clear. The operational dimension is the dimension which is to be obtained from the positioning datum to the surface element. It is the basis of the NC programming and inspection for that operation.

As shown in Figure 2.6, the dimensions in (a) are the dimensions on the blueprint. Dimension 60 ± 0.1 has been machined in another setup. Figures 2.6(b) and (c) show the two most commonly seen operational dimensions found in the CAPP plan literature. On an unmanned NC machine without a touch-trigger probe, this operational dimension is meaningless, since the tool can only be programmed to the machine coordinate system and not on the top of the part, which is a variable. One may think that a touch-trigger probe can solve the problem. Figure 2.6(d) shows the side view of the part. If the two surfaces for dimension 60 are not parallel but the tolerance is still within the specified dimensional tolerance, the variation of dimension 32 should be at least equal to that of dimension 60 and that of the process capability. This example proves that even a CNC machining center

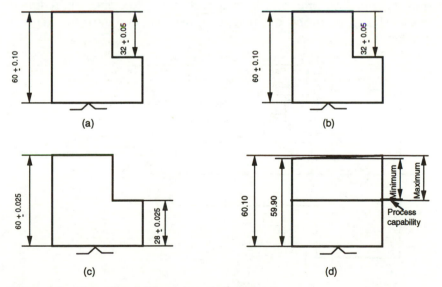

Figure 2.6 Operational dimensions and setup.

with a touch-trigger probe cannot guarantee that the part will be within the specifications, not to mention that the selected CNC machine with touch-trigger probe is an expensive machine and that the operation is time-consuming. To solve the problem, the parallelism of the two faces of dimension 60 should be specified, which may not necessarily be on the blueprint, and it should be less than the tolerance of dimension 32. In Figure 2.6(c), dimension 32 is the resultant dimension from dimension 60 and 28. It can only be obtained if its tolerance is equal or larger than the sum of the tolerances of the other two dimensions, which, in this case have to be compressed. Assume they are 60.000 ± 0.025 and 28.000 ± 0.025 respectively. This example presents an interesting question: if either of the two tolerances for an individual part is out of compressed tolerance (dimension 60 is still within the design tolerance), should it be scrapped or reworked? For all the parts within the designed tolerances, the dimension 28 may vary within 28 ± 0.15. Any value in this range does not necessarily mean that dimension 32 of the part is out of the designed (not operational) tolerance. For example, assume all three surfaces are ideally parallel and the two dimensions are 60.000 and 28.100 respectively. The latter is out of the specified operational tolerance and may be scrapped in operation inspection, but if the part is inspected with the designed specifications, one can find that it is good. From this example we reach the conclusion that if the positioning datum does not coincide with the design datum, some of the tolerances on the part may have to be compressed and false scrap may be produced that may not actually exist. If the two dimensions are machined in one setup, no transformation of datum is involved, and therefore no unnecessary compression of tolerances is required, no touch-trigger probe technology is needed, and no false scrap is produced.

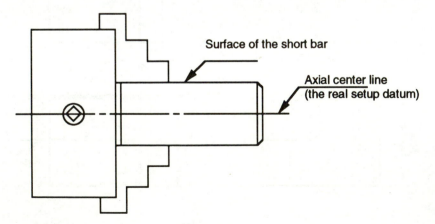

Figure 2.7 An example of a setup datum.

2.5.4 Criteria of process planning

As we have discussed in the previous sections, datums must always be examined and determined before making the process plan. However, sometimes the points, the lines, and the planes to be used as datums may not exist entirely on the part. Instead, the datums are represented by some other concrete features. For example, a short bar stock is clamped by a three-jaw chuck on a lathe in Figure 2.7. In spite of the visible setup datum, namely the external round surface of the short bar, the real setup datum is the axial center line of the short bar. This example indicates that sometimes selecting correct setup datums may actually include selecting the appropriate setup surfaces.

Datum coincidence: Datum coincidence is one of the criteria for selecting setup datums. In order to achieve the datum coincidence, the design datum must usually be selected as a setup datum. If the selected setup datum is not the same as the design datum, processing errors will occur during machining. Since these processing errors are caused by different datums (setup datum and design datum), they are called setup errors.

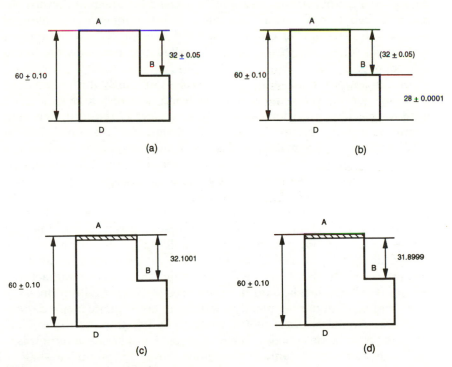

Figure 2.8 An example of a setup error.

Figure 2.8 illustrates a setup error. A workpiece is operated on a CNC milling machine for surface B, as shown in Figure 2.8(a). The CNC machine is not a touch-trigger probe. All cutting dimensions are dependent on the programmed machine. The dimension 60 ± 0.1 (from D to A) is carried over from a previous process. If we use surface D as the setup datum to mill the dimension 32 ± 0.05 (from A to B), the dimension between surfaces D and B has to be converted as shown in Figure 2.8(b). In order to ensure the final result within the design tolerance (32 ± 0.05), an operational tolerance has to be given, while the nominal dimension is converted as 28 mm. Suppose the maximum accuracy of the CNC machine is ±0.0001. In other words, the tightest tolerance we can give to the dimension 28 mm (from D to B) is 28 ± 0.0001. Theoretically, this is a precise machine; however, we may easily see that the maximum dimension achieved from surface B to A is 32.1001, see Figure 2.8(c), and the minimum dimension achieved is 31.8999, see Figure 2.8(d). The results are outside the required design tolerance (32 + 0.05). This example shows that if the setup datum is inadequate, the setup error will be generated from the process.

Similar problems can be encountered when measuring datums are not uniform with design datums. These errors are called measuring errors, as they are caused by unequal measuring datum and setup datum. Some other errors may also be caused by unequal datums. In order to avoid these errors that may directly affect the quality of product, the strategy of datum coincidence must always be applied for datum selection. This strategy is called the principle of datum coincidence.

Selecting appropriate setup datums can significantly improve manufacturing quality and reduce manufacturing cost and lead time. The need for machine precision can also be dramatically reduced. Selecting appropriate setup datums is one of the major concerns while making process plans. As a matter of fact, selection of datums is just the selection of surfaces which can be used for clamping the workpiece. For a stock that has not received any manufacturing processes, a stock surface can only be selected as a setup datum. In this case, the datum, which has not been manufactured before, is called a **black datum**. In the continuing operations of the workpiece, especially for finishing processes such as finish turning, finish milling, finish boring etc., one should always try to select some surfaces which have already been manufactured. In this case the datum, which has been manufactured before, is called a **bright datum**. Sometimes the bright datum does not exist, and so it cannot be directly used for an appropriate setup datum. In this case a surface that will not affect any functions of the original design should be designed, created, and used as a setup datum; this is called an **auxiliary datum**. For instance, while machining a shaft, a center hole is drilled for supporting the shaft on the machine. The center hole, which has no other function except supporting the shaft, is an auxiliary datum.

Another criterion for the selection of datums is the principle of **datum unification**. This strategy indicates that the process planner should try to use the same bright datum to manufacture as many features as possible within a setup. This strategy can avoid generating process errors between these features during manufacturing.

Selection of datums is a critical step in making an optimal process plan. The process planner must be very careful to analyze all of the different datums and select the appropriate one. The principles of datum coincidence and datum unification are strategies to ensure the process's quality. To select appropriate black and bright (especially bright) datums, careful and close attention must be given in examining the relationships between features.

Other than datum selection, another criterion for process planning is equipment selection. Equipment selection for flexible process planning has become a popular research topic under the umbrella of CIM, which requires detailed knowledge of operations research and scheduling. The issue of equipment selection will be discussed in section 10.6.3. However, some criteria for the selection of machine tools must be discussed at this point. While making process plans, the process planner must always try to focus attention on existing equipment within the company. The process planner should try to select a machine which has high productivity and multiple process functions, and may also try to concentrate on as many processes as possible using the high productivity machine. This strategy may reduce the requirements of additional machines, the number of operators, the number of steps in the process, and the route of material handling, as well as the manufacturing lead time. These advantages preclude another strategy of process planning. This strategy is called the principle of **processes concentration**, which means that the process planner should try to concentrate as many processes as possible on a machine within a setup while making process plans.

Some other criteria need to be considered while sequencing the operations. These criteria are as follows:

- All rough processing should be done prior to finish processing
- All major processing should be done prior to minor processing
- All bright datums should be done prior to ordinary features
- All bright datums should be done prior to use

In order to stimulate the application of recent high-tech development, new methodologies, new equipment, new materials, and new strategies should be considered during process planning; for example, CNC machining centers (as a new methodology) and composite materials (as a new kind of material).

2.6 EXAMPLE OF PROCESS PLANNING

In order to provide a clear idea of how to make a process plan in a manufacturing workshop, we are going to introduce two examples based on the previous discussions. Although these two examples cannot enable readers to become qualified process planners, they may be used to learn how to make a process plan for rotational and prismatic machining parts.

2.6.1 Example of a rotational part

For rotational parts, a turning process is almost always required. According to the surface roughness and tolerances, turning processes can be divided into rough turn processes and finish turn processes. Usually, rough turn processes can yield a surface roughness from 1 to 3 (ISO standard), while finish turn processes can yield a surface roughness from 3 to 6. If the surface roughness is higher than 6, a grinding process is usually required. In most cases, a heat treatment is usually assigned before a grinding process is applied, which can increase the surface hardness and ensure the grinding process quality. Center supports at both ends are usually required for a grinding process. If some slots or keyways are designed on the rotational part, a milling process is usually required. The detailed criteria for selecting adequate processes for different surface roughness can be found in Alting (1993) and Kalparjian (1992).

Figure 2.9 illustrates the procedures of process planning for a rotational part, using a shaft part. One hundred pieces are supposed to be ordered for the part. The material is 4140 (a chromium alloy steel, which is a typical material used for shaft parts). The corresponding procedures and illustrations of the process planning are as follows:

NAME OF THE PART: SHAFT
MATERIAL: 4140 ALLOY STEEL
STOCK: BRIGHT BAR

Interpretation of design data

This part is a single axis rotational part. The three outer diameters have tolerances needing the grinding process. There is a keyway on the shaft that needs the milling process. The total runout of 20 mm diameter uses two 18 mm diameters as datum elements. The maximum length is 150 mm and the maximum diameter is 25 mm. Center supports at both ends are necessary from the viewpoint of tolerance and manufacturing. Turning (on a lathe), milling and grinding processes are selected for manufacturing this part.

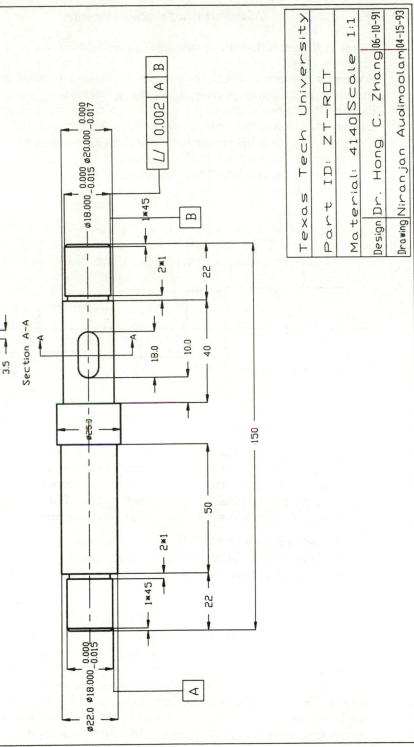

Figure 2.9 An example of a rotational part.

Setup 1: Rough turn the left side of the part
Machine: lathe
Fixture: 3-jaw chuck
Reference point (axial): 3 jaws of the chuck
Extension: 92 mm
Operations: 1. face the end
            2. drill the center hole and center support it
            3. rough turn 25 mm diameter
            4. rough turn 22 mm diameter
            5. rough turn 18 mm diameter

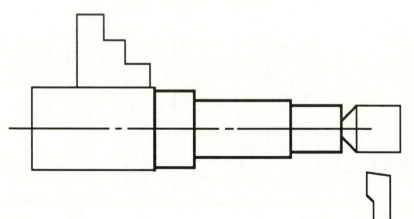

Figure 2.10(a)

Setup 2: Rough turn the right side of the part
Machine: lathe
Fixture: 3-jaw chuck
Reference point (axial): 3 jaws of the chuck
Datum: 22 mm diameter and the shoulder between it and 25 mm
       diameter
Operations: 1. face the end
            2. drill the center hole and center support it
            3. rough turn 20 mm diameter
            4. rough turn 18 mm diameter

Setup 3: Finish turn the left side of the part
Machine: lathe
Fixture: center support at both ends and driving accessory
Datum: two center holes

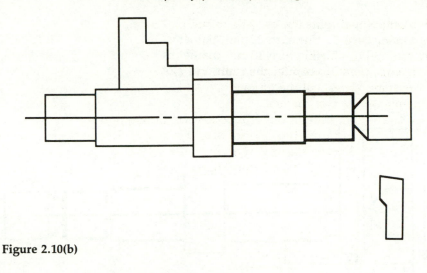

Figure 2.10(b)

Operations: 1. finish turn 25 mm diameter to dimension
2. finish turn 22 mm diameter to dimension
3. fine turn 18 mm diameter
4. chamfer the left end 1X45
5. cut the groove

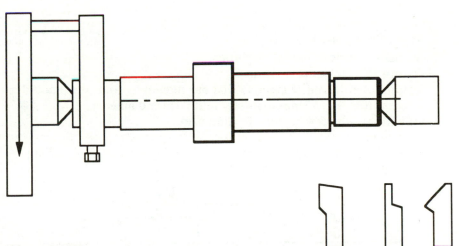

Figure 2.10(c)

Setup 4: Finish turn the right side of the part
Machine: lathe
Fixture: center support at both ends and driving accessory

Datum: two center holes
Operations: 1. fine turn 20 mm diameter
2. fine turn 18 mm diameter
3. chamfer the right end 1X45
4. cut the groove

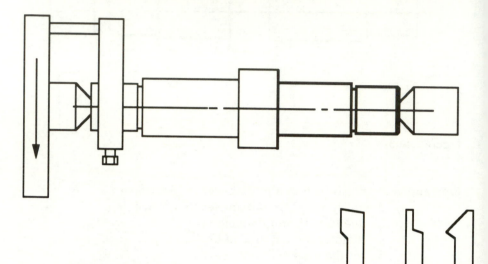

Figure 2.10(d)

Setup 5: Mill the keyway
Machine: vertical milling machine
Fixture: V-block and U-clamp
Datum: left 18 mm diameter, 20 mm diameter, and the shoulder
between 25 mm diameter and 20 mm diameter
Operation: mill the keyway to dimension

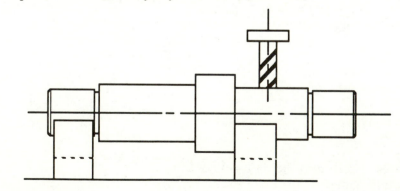

Figure 2.10(e)

Heat Treatment

Setup 6: Grind left 18 mm diameter
Machine: cylindrical grinding machine
Fixture: center support at both ends and driving accessory
Datum: two center holes
Operations: grind the 18 mm diameter to dimension

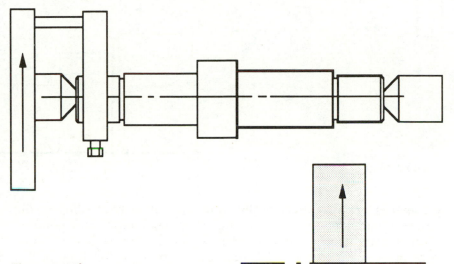

Figure 2.10(f)

Setup 7: Grind right 18 mm and 20 mm diameters
Machine: cylindrical grinding machine
Fixture: center support at both ends and driving accessory
Datum: two center holes
Operations: 1. grind the 18 mm diameter to dimension
2. grind the 20 mm diameter to dimension

Inspection

The process plan for the part is shown in Figure 2.11.

2.6.2 Example of a prismatic part

In general, process planning for a prismatic part is usually more complicated than for a rotational part because the geometrical shapes of

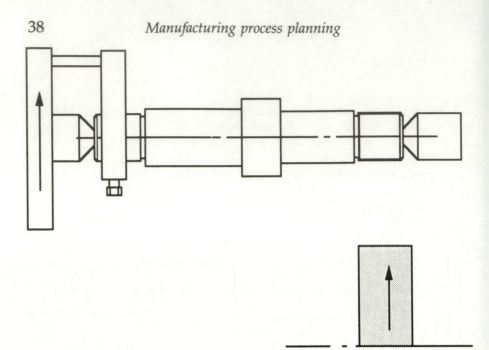

Figure 2.10(g)

prismatic parts vary dramatically from each other. According to the different attributes, milling, drilling, shaping, boring, and grinding processes are usually required. Here again, the fundamental knowledge of selection of machining processes is beyond the scope of this book; interested readers can refer to Alting (1993) and Kalparjian (1992).

To provide a general idea of process planning for prismatic parts, we select a gear box block as an example, as shown in Figure 2.12. In order to simplify the discussion, only major dimensions relevant to datum selection for the gear box block are shown in Figure 2.12. The procedure of process planning is focused only on the major processes and datum selection. The corresponding procedures and illustrations of the process planning for the gear box block are as follows:

NAME OF THE PART: GEAR BOX BLOCK
MATERIAL: GREY IRON CASTING

Interpretation of design data

The part has some plane surfaces to be machined, some big holes to hold bearings for shafts, and some small holes for assembling the part with other parts. Both the dimensions of, and the distances between, the big holes are accurate. Considering both efficiency and accuracy, we select a milling machine for the plane surfaces, a boring machine for the big holes, and a drilling machine for the small holes.

PROCESS PLAN

Part ID: SHAFT (Example 2.2)   Planner: J. Mei

P#	O#	Description		Tools	Setup T.	Mach. T
10	(Stock preparation) -				
		Raw material storage				
	10	4140 alloy steel bright bar	D=28 mm L=154 mm			
20	(Rough turn 1) -				
		CNC lathe A			0.50 h	0.13 h
	10	Clamp part with ext.	L=92 mm	3-jaw chuck		
	20	Facing section	depth=2.00 mm	Facer		
	30	Drill center	depth=5.00 mm	Cent. Driller		
	40	Tailstock & Center supporting		Center		
	50	Rough turn diameter	D=25.00 mm	Sidebit		4.50 m
			X= 0.00 mm L=89.00 mm			
	60	Rough turn diameter	D=22.00 mm	Sidebit		2.50 m
			X= 0.00 mm L=72.00 mm			
	70	Rough turn diameter	D=18.00 mm	Sidebit		1.00 m
			X= 0.00 mm L=22.00 mm			
30	(Rough turn 2) -				
		CNC lathe B			0.50 h	0.06 h
	10	Clamp part on diameter D=22 mm				
		using the shoulder (ext. L=80 mm)		3-jaw chuck		
	20	Facing section	depth=2.00 mm	Facer		
	30	Drill center	depth=5.00 mm	Cent. Driller		
	40	Tailstock & Center supporting		Center		
	50	Rough turn diameter	D=20.00 mm	Sidebit		2.50 m
			X= 0.00 mm L=62.00 mm			
	60	Rough turn diameter	D=18.00 mm	Sidebit		1.00 m
			X= 0.00 mm L=22.00 mm			
40	(Finish turn 1) -				
		CNC lathe C			0.50 h	0.95 h
	10	Double center supporting		Cent. & Tail.		
	20	Finish turn diameter	D=25.00 mm	Finebit		1.00 m
			X=72.00 mm L=16.00 mm			
	30	Finish turn diameter	D=22.00 mm	Finebit		3.00 m
			X=22.00 mm L=50.00 mm			
	40	Fine turn diameter	D=18.00 mm	finebit		1.00 m
			X= 0.00 mm L=22.00 mm			
	50	Turn left chamfer	A=45 deg.	Sidebit		0.15 m
			X= 0.00 mm W= 1.00 mm			
	60	Turn square groove	B= 2.00 mm	Grosqubit		0.50 m
		X=22.00 mm	D= 1.00 mm			
50	(Finish turn 2) -				
		CNC lathe D			0.50 h	0.07h
	10	Double center supporting		Cent. & Tail.		
	20	Fine turn diameter	D=20.00 mm	Finebit		2.50 m
			X=22.00 mm L=40.00 mm			
	30	Fine turn diameter	D=18.00 mm	Finebit		1.00 m
			X= 0.00 mm L=22.00 mm			
	40	Turn right chamfer	A=45 deg.	Sidebit		0.15 m
			X= 0.00 mm W= 1.00 mm			
	50	Turn square groove	B= 2.00 mm	Grosqubit		0.50 m
		X=22.00 mm	D= 1.00 mm			
60	(Milling keyway) -				
		CNC milling machine A			0.10 h	0.40 h
	10	Mill keyway	B= 8.00 mm	Endmill A		2.40 m
			D= 5.00 mm L=18.00 mm			
70	(Heat treatment) -				
		Heat-treatment				
80	(Grinding 1) -				
		CNC cylindrical grinding machine A			0.05 h	0.07 h
	10	Double center supporting		Cent. & Tail.		
	20	Grind diameter	D=18.00 mm	Sandwheel		2.00 m
90	(Grinding 2) -				
		CNC cylindrical grinding machine B				
	10	Double center supporting		Cent. & Tail.		
	20	Grind diameter	D=18.00 mm	Sandwheel		2.00 m
	30	Grind diameter	D=20.00 mm	Sandwheel		

Figure 2.11 Process plan for the rotational part.

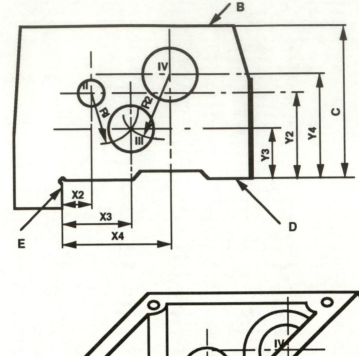

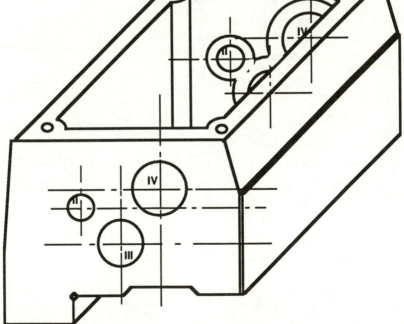

Figure 2.12 An example of a prismatic part.

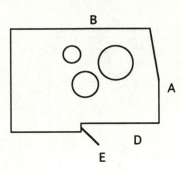

Figure 2.13(a)

Setup 1: Face mill surface B
Machine: milling machine
Fixture: machine table, blocks and clamps
Datum: surface D
Operations: face mill surface B

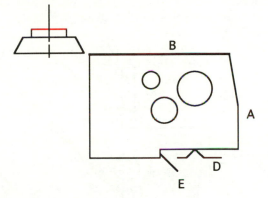

Figure 2.13(b)

Setup 2: Mill surface D and surface E
Machine: milling machine
Fixture: machine table and clamps
Datum: surface B and surface A
Operations: 1. face mill surface D
2. end mill surface E

Setup 3: Bore the big holes
Machine: boring machine
Fixture: machine table, blocks and clamps

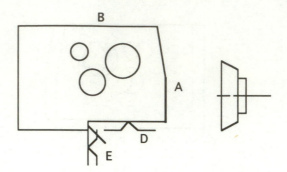

Figure 2.13(c)

Datum: surface D and surface E
Operations: 1. rough bore the holes and face (countersink) side
surfaces
2. fine bore the holes and face (countersink) side
surfaces
3. finish bore the holes to dimension

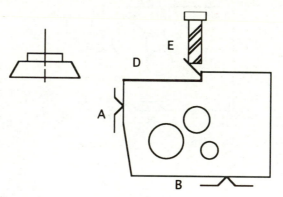

Figure 2.13(d)

Setup 4: Drill and tap the small holes
Machine: drilling machine
Fixture: machine table, blocks and clamps
Datum: surface D and surface E
Operations: 1. drill the small holes
2. tap the holes

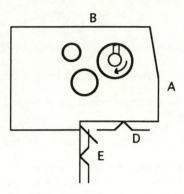

Figure 2.13(e)

Inspection

The process plan for the gear box block part is shown in Figure 2.14.

```
--------------------------------------------------------------------
                                                      04-09-1993
                          PROCESS  PLAN
Part ID: GEAR BOX BLOCK (Example 2.3)                 Planner: J. Mei
--------------------------------------------------------------------
 P#   O#   Description                   Tools         Setup T. Mach. T.
--------------------------------------------------------------------

 10 (Stock preparation: Casting) - - - - - - - - - - - - - - - - - - - - - -
           Cast part shop
      10   Grey cast iron casting
 20 (Milling 1) - - - - - - - - - - - - - - - - - - - - - - - - - - - - -
           CNC milling machine A                      0.50 h  0.67h
      10   Clamp part on machine table   Block & Clamp
               Datum: D E
      20   Face mill surface B           Facemiller              30 m
               Y = C
      30   Face mill surface A           Facemiller              10 m
 30 (Milling 2) - - - - - - - - - - - - - - - - - - - - - - - - - - - -
           CNC milling machine B                      0.50 h  0.50 h
      10   Clamp part on machine table   Block & clamp
               Datum: B A
      20   Face mill surface D           Facemiller              20 m
               Y = C
      30   End mill surface E            Endmill                 10 m
 40 (Boring)    - - - - - - - - - - - - - - - - - - - - - - - - - - -
           Boring Machine A                           0.50 h  2.00 h
      10   Clamp part on machine table   Block & clamp
               Datum: D E
      20   Rough bore and countersink holes  Borebit             30 m
      30   Fine bore      and countersink holes  Borebit          40 m
      40   Finish bore holes             Borebit                 50 m
 50 (Drilling) - - - - - - - - - - - - - - - - - - - - - - - - - - -
           Drilling machine A                         0.50 h  1.33 h
      10   Clamp part on machine table   Block & clamp
               Datum: D E
      20   Drill small holes             Drillbit                40 m
      30   Tap small holes               Tapper                  40 m
 60 (Inspection)  - - - - - - - - - - - - - - - - - - - - - - - - - - -
           Inspection station
      10   Inspection
--------------------------------------------------------------------
```

Figure 2.14 Process plan for the prismatic part.

REFERENCES

Allen, D.K. (1987) An introduction to Computer-Aided Process Planning. *CIM Review*, Fall 1987.

Alting, L. (1982) *Manufacturing Engineering Processes*, Marcel Dekker Inc., New York, p. 367.

Alting, L. (1993) *Manufacturing Engineering Processes*, Second edition, Marcel Dekker Inc., New York.

Alting, L. and Zhang, H. (1989) Computer Aided Process Planning: The state-of-the-art survey. *International Journal of Production Research*, **27**(4), pp. 553–85.

Kalparjian, S. (1982) *Manufacturing Engineering and Technology*, Second edition, Addison-Wesley Publishing Company, USA, p. 1258.

Tuttle, H.C. (1983) Breaking the wall between design and manufacturing. *Production*, May, pp. 63–66.

CHAPTER 3

Impact of Computer Integrated Manufacturing

3.1 INTRODUCTION

Computer Integrated Manufacturing (CIM) has been the keyword in many papers, textbooks, technical conferences, and seminars in the last few years. The aim of CIM is to achieve completely automated production, also referred to as 'the factory of the future', by integrating the isolated automated production islands which have been the key solutions to the problems existing in the discrete manufacturing industry. These isolated automated production islands include Numerical Control (NC), Distributed Numerical Control (DNC), Computerized Numerical Control (CNC), Material Requirement Planning (MRP I), Manufacturing Resource Planning (MRP II), Computer Aided Design (CAD), Computer Aided Process Planning (CAPP), Computer Aided Manufacturing (CAM), automated storage, computer controlled material handling equipment, Automated Guided Vehicles (AGV), robotics, etc.

L. Alting has stated, 'Today CIM is looked upon as a strategy for planning, implementation and integration of many functions in a manufacturing company by the use of computers. These functions include all the different departments, and all levels in the factory. In other words, CIM, as a strategy of the factory of the future, has a great impact on the whole manufacturing enterprise, both in the daily operation of the factory, and in changes performed in the production environment (new machines, new parts, new operators, new processes, new methodologies, new technologies, new equipment, new organization of enterprises, new structure of company, new policies of company etc.)' (Alting *et al.* 1987).

In this chapter, the elements which play important roles for integrated production, such as Group Technology (GT), CAD, CAM, and Manufacturing Database (MD), will be discussed. The final integrated system will then be described.

3.2　COMPUTER AIDED DESIGN

3.2.1　Introduction to Engineering Design

Engineering design is the logical step prior to manufacturing process planning. In general, engineering design can be related to different fields. According to the objective of the design task, engineering design can be specified as mechanical engineering design, electrical or electronics engineering design, civil engineering design, chemical engineering design, agricultural engineering design, or municipal and architectural design. All of these engineering designs usually consist of five basic steps:

1. conception of design
2. synthesis of design
3. analysis of design
4. evaluation of design
5. representation of design

The conception of the design is usually generated by a discussion among design engineers, sometimes with other participators such as salesmen, whose participation is based on the demands for the new product and information regarding trends from the market. This initial design is usually very aggregate. The emphasis of this design stage is focused on the physical and functional characteristics, cost, quality, and operation performance of the new product, and comparisons with the old product. This design stage normally contains the general components of the product but lacks specific details. The conception of design is usually carried out in the form of a sketch.

　Synthesis and analysis of the design are closely related and highly iterative in the design process. A certain component or subsystem of the overall system is conceptualized by the designer, subjected to analysis, improved through an analysis procedure, and redesigned. The process is repeated until the design has been optimized within the constraints imposed on the designer. The components and subsystems are synthesized into the final overall system in a similar iterative manner.

　Evaluation of the design is concerned with measuring the design against the specifications established in the problem definition phase. This evaluation often requires the fabrication and testing of a prototype to assess operating performance, quality, reliability, and other criteria. The final phase in the design process is the presentation of the design. It includes documentation of the design by means of drawings, material specifications, assembly lists, etc. Essentially, the documentation requires that a design database be created. In general, computer aided design software can be used beneficially in the five phases of the

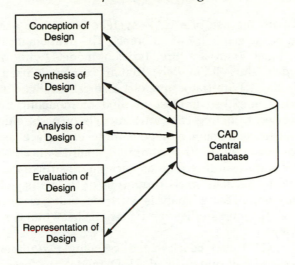

Figure 3.1 Computer aided design phases.

design process as illustrated in Figure 3.1. However, different systems may have different advantages and disadvantages with respect to these steps. Computer aided graphics may be the common function that all CAD systems can possess.

Engineering design has traditionally been accomplished on drawing boards and then documented in the form of a detailed engineering drawing. Mechanical design includes the drawing of the complete product as well as its components and subassemblies, and the tools and fixtures required to manufacture the product. There are several methods available to represent an engineering drawing. The conventional method is drafting on paper with pen or pencil. Manual drafting is tedious and requires a tremendous amount of patience and time. Recently, computer aided drafting systems have been implemented to improve drafting efficiency. The major objective of these systems is to assist the draftsmen with tedious drawing and redrawing. Partially compiled or completed drawings from a graphics tablet or screen can be stored in a computer and retrieved when needed.

3.2.2 Computer Aided Design

Computer Aided Design (CAD) can be defined as using computers to aid the engineering design process by means of effectively creating, modifying, or documenting the part's geometrical modeling. CAD is most commonly associated with the use of an interactive computer graphics system. The object of the engineering design is stored and represented in the form of geometric models. Geometric modeling is concerned with the use of a CAD system to develop a mathematical description of the geometry of an object. The mathematical description is called a model. There are three types of models (wire-frame models, surface models, and solid models), that are commonly used to represent a physical object. Wire-frame models, also called edge-vertex or stick-figure models, are the simplest method of modeling and are most commonly used to define computer models of parts. Surface models may be constructed using a large variety of surface features. Solid models are recorded in the computer mathematically as volumes bounded by surfaces rather than as stick-figure structures. As a result, it is possible to calculate mass properties of the parts, which is often required for engineering analysis such as finite element methods, kinematic or dynamic studies, and mass or heat transfer for interference checking.

Models in CAD also be classified as being two-dimensional (2D) models, two-and-half-dimensional (2½D) models, or three-dimensional (3D) models. A 2D model represents a flat part and a 3D model provides representation of a generalized part shape. A 2½D model can be used to represent a part of constant section with no side-wall details. The major advantage of a 2½D model is that it gives a certain amount of 3D information about a part without the need to create the database of a full 3D model.

After a particular design alternative has been developed, some form of engineering analysis must often be performed as a part of the design process. The analysis may take the form of stress – strain calculations, heat transfer analysis, dynamic simulation etc. Some examples of the software typically offered on CAD systems are mass properties and Finite Element Method (FEM) analysis. Mass properties analysis involves the computation of such features of a solid object as its volume, surface area, weight, and center of gravity. FEM analysis is available on most CAD systems to aid in heat transfer, stress–strain analysis, dynamic characteristics, and other engineering computations. Presently, many CAD systems can automatically generate the 2D or 3D FEM meshes which are essential to FEM analysis.

As a matter of fact, development of CAD systems is now quite mature. However, considering the interface between CAD and CAPP, many problems still remain. The main problem is transformation of geometrical models or, more strictly, geometrical model representations, from CAD to CAPP (Alting and Zhang, 1989). For instance, in the simplest 2D form, Initial Graphics Exchange Specification (IGES) can represent an engineering drawing, but items such as dimensions can be represented in different ways. Also different drafting systems use different technologies to group lines into profiles. As a result, there appear to be major problems in using IGES to transfer data between different systems. In 3D, the problems are worse because many ways of sorting surface and space curves are incompatible. Some other attempts, such as the approach of Boundary Representation (B-Rep) and the approach of Constructive Solid Geometry (CSG) trees in which the cavities are recognized from the special relationships between the primitive volumes, do not provide any semantic information which could be associated with the machined volumes and are based on local information. Nevertheless, great efforts have been made in this area, and many approaches have been provided to interface CAPP with CAD (Chang *et al.*, 1988; Jakubowski, 1982; Kyprianou, 1983; Henderson, 1984; Kung, 1984; Bjorke, 1984; Joshi and Chang, 1988).

3.3 COMPUTER AIDED MANUFACTURING

The term Computer Aided Manufacturing (CAM) covers many areas from information processing and decision making to manufacturing and machining, which makes giving a single definition for CAM extremely difficult. D. Kochan gave a very fitting definition for CAM, with its diversity and wide range of use, in his book, 'CAM – discrete manufacturing is defined as computer-aided preparation of manufacturing including decision-making, process and operational planning, software design techniques, and artificial intelligence, and manufacturing with different types of automation (NC – machine, NC – machine centers, NC – machining cells, NC – flexible manufacturing systems), and different types of realization (CNC – single unit technology, DNC – group technology)' (Kochan, 1986).

Since CAM has such a wide range of uses, a better way to look at CAM is through CAM technologies. It is not possible to cover all CAM technologies in one book, much less in one section of a chapter, but some of the CAM technologies are covered in different sections of this chapter. The CAM technologies covered are group technology, manufacturing database, automated and tolerancing. As a matter of fact, the whole theme of this book lies in the area covered under the umbrella of CAM. Figure 3.2 illustrates the general scope of CAM.

The essential role of the computer in the production function is to capture and process the data relating to a large number of transactions which continuously take place in different departments of the company. The initial research activity for CAM was Numerical Control (NC) for machine tools at the Massachusetts Institute of Technology (MIT) in 1953. The first programming language was Automatically Programming Tools (APT) created at MIT, and it was the pattern for many further developments. Currently, many manufacturing functions have been addressed by CAM including the following:

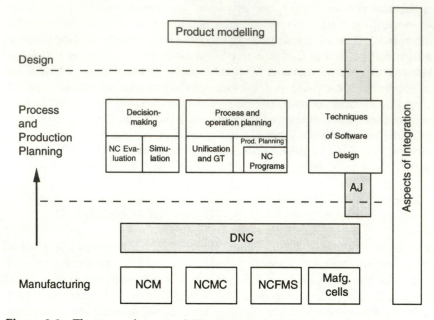

Figure 3.2 The general scope of CAM (Kochan, 1986).

- Numerical Control (NC)
- Computer Numerical Control (CNC)
- Direct Numerical Control (DNC)
- Computer controlled conveyor systems
- Computer controlled machining process
- Computer aided process monitoring
- Computer aided fixturing design
- Computer aided tooling design
- Computer aided tolerancing analysis
- Computer aided cost estimating
- Material Requirement Planning (MRP)
- Computer Aided Process Planning (CAPP)
- Computerized machinability data systems
- Manufacturing Resources Planning (MRP II)
- Computer aided decision support systems
- Development of work standards
- Computer aided line balancing
- Production and inventory planning
- Computer aided scheduling
- Computer aided quality control
- Computer aided inspection

Computer numerical control

Numerical control (NC) is a form of programmable automation in which the processing equipment is controlled by means of numbers, letters, and other symbols. The numbers, letters, and symbols are coded in an appropriate format to define a program of instructions for a particular workpiece or job. When the job changes, the program of instructions for the particular workpiece changes. The capability to change the program is what makes NC suitable for low-volume and medium-volume production, and it is much easier to write new programs than to make major alterations to the processing equipment. (Groover, 1987).

The principle of numerical control was first applied to the milling process, and then later to the turning process, flame cutting, drilling, and grinding. NC technology is now used more and more for other manufacturing processes, such as forming (fine forging, ringing, rolling, etc.), engraving, and laser cutting.

The current NC equipment is relatively more mature. Many machines possess multiple processing functions, such as milling centers which can perform vertical and horizontal milling, drilling, boring, reaming, slotting, shaping, and turning processes. Of course, with a high capacity automated tooling library, CNC machines' functions can be considerably more abundant.

Programmable logic controller

Programmable logic controllers (PLCs) are widely used in computer aided manufacturing. Actually, PLCs are used in virtually every segment of industry where automation is required. PLCs represent one of the faster growing segments of the electronics industry. Since their inception, PLCs have proved to be the salvation of many manufacturing plans which previously relied on electro-mechanical control systems. A PLC is a solid-state device designed to perform logic functions previously accomplished by electro-mechanical relays. The design of most PLCs is similar to that of a computer. Basically, the PLC is an assembly of solid-state digital logic elements designed to make logical decisions and provide outputs. Programmable logic controllers are used for the control and operation of manufacturing process equipment and machinery.

Computer aided material handling

Material handling (MH) is a very important factor in how efficiently a workshop or company can be operated. An efficient MH system will help reduce waiting time, and it may even help increase safety or the effectiveness of the entire manufacturing process.

Cabbert and Brown indicated that as much as 60% of the total production cost may be accounted for by material handling (Laing *et al.*, 1989). It is also evidenced that most discrete manufacturing products spend 90% of their manufacturing lead time on the duration of material handling and storage. With MH accounting for such a large amount of the total production cost, it is obvious that reducing the amount of time a product is handled will dramatically reduce production costs. One way of helping reduce these costs is by using computers to do some material handling.

There is a great variety of material handling equipment available commercially and there are many types of MH approaches used today. One of these approaches is to use a computer database to store listings of MH equipment and the user's input of factor values. The computer takes the user's required level of, and preferred importance for, each criterion, and the feasible MH equipment for the task at hand, and produces a category of equipment from which the user can choose the proper type or piece of MH equipment (Laing *et al.*, 1989).

Computer monitoring and diagnostics for manufacturing processes

In a computer monitoring and diagnostic system, the aim of monitoring is to detect failures, while the aim of diagnostics includes fault localization and identification. Both monitoring and diagnostics should appear at all levels of the control–monitoring hierarchy.

There are some essential requirements that almost every monitoring and diagnostic system should possess. Some of the requirements for a monitoring system are: (1) the ability to measure and process relatively numerous analogue and digital signals; (2) the capability of profound preprocessing of measured signals, including statistical and frequency based analysis; (3) the ability for complex, multi-parameter decisions; (4) modular, extendable, reconfigurable structure; (5) programmability in all functions; and (6) standardized bi-directional software/hardware interfaces to the CNC/DNC controllers. Some of the requirements for a diagnostic system are: (1) the system should easily provide knowledge about the causal interrelationship when faults arise, to enable even workers who are not well acquainted with the process to localize faults; (2) the consequences of faults should be readily available in the system so that the severity of a given fault for the further production process can be estimated; (3) the user should have the possibility of repairing the fault alone, i.e. repair instructions should be available to the users in a suitable form; (4) the operation of the expert system should be possible by employees who have no previous experience with computers; and (5) after a short training period, the system should be maintained by the employees running the facility so that the presence of expert engineers is no longer necessary (Spur *et al.*, 1989a,b).

L. Monostori *et al.* stated in their paper on monitoring and diagnostic systems that there are three major types of M/D systems that can be classified by their place and function in the manufacturing system. These M/D systems are: (1) autonomous subsystem monitoring, which gets only messages containing environment or condition descriptions from upper levels of control, and supplies all of the elements of the monitoring process with instructions, parameters, or settings needed for measuring, processing, classification, and intervention; (2) complementary subsystem monitoring, which undertakes only the task of measuring and processing and passes classification and intervention to the system level; and (3) semi-autonomous monitoring, which performs only simple, quick monitoring functions autonomously on its own level, and turns to upper levels in the case of sophisticated classification and intervention tasks. (Monostori *et al.*, 1989).

The ideal computer monitoring and diagnostic system can be summed up as being a system that can be used during the absence of the human expert, for example, when the expert is on vacation, during breaks, or if a company wants to have three shifts with few people on the third shift.

3.4 MANUFACTURING DATABASE

A database system is basically a computerized record keeping system. In other words, it is a system whose overall purpose is to maintain information and to make that information available on demand. The information can be anything that seems to be of significance to the individual or organization that the system is intended to serve. The major user of the database system will be given equipment to perform a variety of operations, including

- Adding new empty files to the database
- Inserting new data into existing files
- Retrieving data from existing files
- Updating data in existing files
- Deleting data from existing files
- Removing existing files permanently from the database (Date, 1986).

3.4.1 Introduction to database systems

In general, a database architecture can be illustrated as in Figure 3.3. The architecture is divided into the following three levels:

- **The internal level**, which is the one closest to physical storage, is the one concerned with the way the data is actually stored

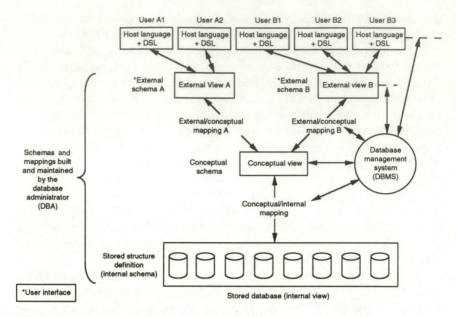

Figure 3.3 A general architecture of a database.

- **The external level**, which is the one closest to the users, is the one concerned with the way the data is viewed by individual users
- **The conceptual level** is a level of indirection between the other two

Today, there are many different types of databases. While many other types of database are being developed, relational databases dominate the application of databases. As a matter of fact, almost all of the database systems developed over the past few years are relational databases. The relational database model was formally introduced by Codd (Codd, 1970) and since then has been through a series of writings.

A relational database system is a system in which the data is perceived by the user as tables (and nothing but tables), and provides operators that, at the user's disposal, generate new tables from the old ones. An explicit example can give a better understanding of the issue. One operator will extract a subset of the rows of a given table, and another operator will extract a subset of columns. A row subset and a column subset of a table may both be regarded as tables themselves. Date has given a precise definition of the relational database: 'a relational database is a database that is perceived by the user as a collection of time-varying, normalized relations of assorted degrees.' (Date, 1986). The idea of the relational model applies at the external and conceptual levels of the system but not at the internal level.

With respect to relational databases, some other types of databases can also be categorized according to the data structures and operators that are presented to the user. The four major categories are relational, inverted list, hierarchic, and network.

An inverted list database is similar to a relational database, but a relational database is at a low level of abstraction, at which the stored tables themselves, and also certain access paths to those stored tables, are directly visible to the user. In general, the inverted list database has the following three significant differences from the relational database:

- The rows of an inverted list table are considered to be ordered in some physical sequence
- An ordering may also be defined for the total database in the form of a database sequence
- For a given table, any number of search keys can be defined in an arbitrary field or combination of Fields

A hierarchic database consists of an ordered set of trees; more precisely, an ordered set consisting of multiple occurrences of a single type of tree. The data in the hierarchic database is represented to the user in the form of a set of tree structures, and the operators are provided with means for traversing hierarchy paths up and down the trees. A tree type consists of a single root record type together with an ordered set of zero or more dependent (lower level) sub-tree types. A sub-tree type also consists of a single record type, which is the root of the sub-tree type together with an ordered set of zero or lower level dependent sub-tree types. The entire tree type thus consists of a hierarchical arrangement of record types. In addition, of course, record types are made up of field types in the usual way. As an example, Figure 3.4 illustrates a hierarchical geometrical configuration database.

A network database structure can be regarded as an extended form of the hierarchic data structure. The principal distinction between the two is as follows: in a hierarchic structure, a child record has exactly one parent, and in a network structure, a child record can have any number of parents (possibly zero). A network database consists of two sets, a set of records and a set of links. More accurately, it consists of a set of multiple occurrences of each of several types of records, together with a set of multiple occurrences of each of several types of link. Each link type involves two record types, a parent record type and a child record type. Each occurrence of a given link type consists of a single occurrence of the parent record type together with an ordered set of multiple occurrences of the child record type. As an example, Figure 3.5 illustrates a network type database for machine tools.

As mentioned previously, the field of database research and development is certainly growing. The application of databases becomes important in manufacturing enterprises, especially for companies which are searching for integrated production solutions. As a matter of fact,

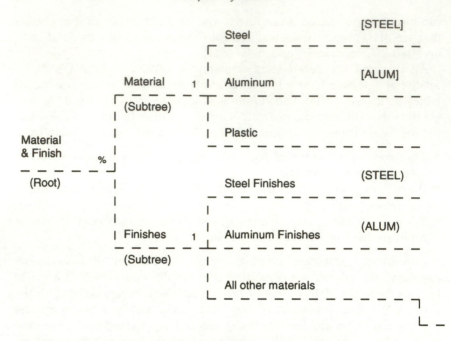

Figure 3.4 A hierarchical geometrical configuration database.

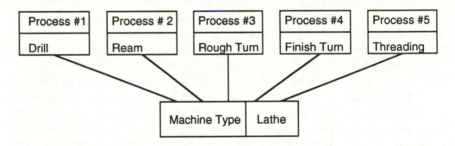

Figure 3.5 A network type machine tools database.

the application of distributed database systems results in more potential advantages in integrated production, especially in a concurrent engineering environment.

A distributed system is any system involving multiple sites connected in some kind of communication network, in which a user at any site can access data stored at any site. Each site, in turn, can be thought of as a database system in its own right. It has its own database, its own DBA function, its own terminals and users, and its own local storage and CPU. An example of a distributed system is shown in Figure 3.6. The main objective of a distributed system is to provide location

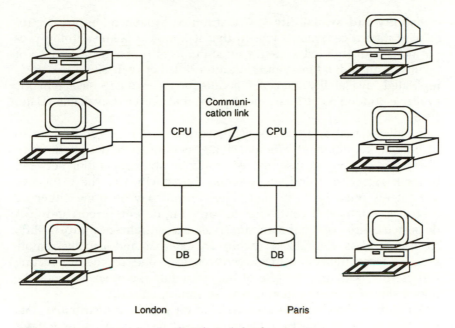

London   Paris

Figure 3.6 An example of a distributed database system.

transparency, i.e. users need not know at which site any given piece of data is stored, but should be able to behave as if the entire database were stored at their own local site (Draffan and Poole, 1980).

Advantages of distributed systems

Local autonomy – The enterprise served by the system is certainly distributed into divisions or departments, or into plants, factories, etc. Distributing the system allows individual groups within the enterprise to exercise local control over their own data and generally makes them less dependent on some remote data processing center. At the same time, distributing the system allows those local groups to access data at other locations when necessary.

Improvement in efficiency – The small computers found within the distributed systems are normally dedicated, at least for a period of time, to one task only. Therefore, these computers have a much higher efficiency in terms of cost per unit of work done than large multi-programming, centralized systems.

Flexibility – If the information processing requirements of the organization change, then data can be dynamically moved or replicated, or existing replicas can be deleted.

Reliability and availability – A distributed system offers greater reliability than a centralized one in that it is not an all-or-nothing pro-
position. A distributed system can continue to function in the face of
failure of an individual communication link between sites. If data is
replicated, availability increases because a given data object remains
available so long as at least one copy of that object is available (Date,
1986).

If a distributed approach is valid, economic, and a good match to
operational requirements, then such a system must be designed. Since
distributed computing systems necessarily involve cooperation between
two different computing subsystems, the activities of the respective
subsystems must be controlled. The control may be done either by
means of a dedicated control center or using decentralized control. A
dedicated control center is sometimes called a centralized control facility,
since control is centralized within an external and usually remote
system. Decentralized control involves any one of the interacting
subsystems assuming the role of master, while the other assumes the
role of slave for the duration of any session or liaison.

A number of techniques are available for handling distributed data
which must be shared by two or more subsystems. Most of today's
distributed systems work with other subsystems within the same
company or organization. As horizons extend and it becomes desirable
for distributed systems to work with other subsystems owned by different companies, standards will come into sharp focus. For example,
if two subsystems are proposing to link, then the subsystems must
determine that such a link is permissible and useful before the actual
link is established. Means which allow one subsystem to interrogate
the capabilities or resources of another are a necessary part of such
standards.

The manufacturing database has played an important role in Integrated Production (IP). In the manufacturing environment, it is usually
necessary to maintain a very large volume of data which is accessible
to, and used by, a number of production departments within the
company. In the early stages of the application of computation in
industry, this data was organized in the form of a number of simple
files, or preferably as a number of linked files using the direct access
capability of the discs, so that the data could be retrieved in the
required format. The usual practice was to store such data, particularly
the large files, on magnetic tapes (Kochhar, 1979). This was probably
the initial description of the prototype for the manufacturing database.

3.4.2   Databases in manufacturing engineering

In order to manage a company effectively, manufacturers use and
store large amounts of information on designs, inventory, outstanding

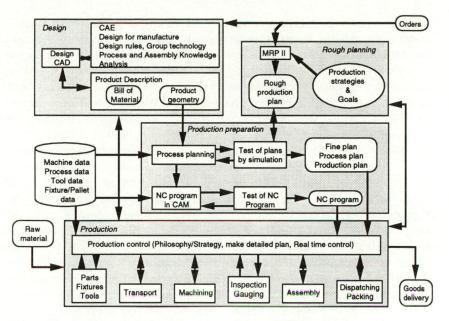

Figure 3.7 Interconnected production activities (Bilberg, 1989).

orders, capabilities of different machines, personnel, and costs of raw materials, to name only a few. These large amounts of data are not static and require constant updating, as correct decisions can only be made by using up-to-date information which reflects the current situation of the company. In some manufacturing businesses, the current work flow is often delayed by the tasks of creating paper, moving paper, reforming and transcribing data, and a variety of checking, validating, searching for information, liaison, meetings, and administration.

The advent of computers offers a means for collecting and storing information on the operations of the company in databases. The ability to access these databases quickly and effectively is a powerful management tool. In addition to saving labor, computerized manufacturing information systems bring more flexible and more widespread access to company information and thus assist staff at all levels in making informed decisions. Figure 3.7 shows an information flow in interconnected functions of all processes of a manufacturing enterprise.

Generally speaking, engineering databases are more complex than those typically found in business. Therefore, these databases require management strategies geared to the engineering process and to the data produced. Engineering database management involves controlling information on three levels and thus requires tools for three types of users: engineers, managers, and software developers.

Engineers are primarily interested in developing models and moving them from design to manufacturing. To do this efficiently requires tools that can be used to store and manipulate data at the workstation or terminal. The tools that engineering managers require are based on database managers and include such functions as revision, access, and release control, as well as file management. Software developers require tools that allow extensions of the basic DBMS for customizing the database to the user's business requirements.

Unlike traditional business databases, which are composed of homogeneous records, engineering databases are usually made of files of varying lengths that include some combinations of geometric and alphanumeric data. Therefore, database management schemes designed to keep track of accounting, records, etc., cannot really be adapted to systems where some files might include model geometry, some analysis data, and some bills of materials.

Engineering database managers should handle engineering change control and documentation, configuration data, and release procedures. Also, since engineering organizations tend to be as diverse as the data they produce, engineering DBMSs should provide users with several layers of customization.

First, engineering DBMSs should allow customization of applications. For instance, they should allow users to choose the method used to retrieve data. Second, engineering DBMSs should allow users to customize company approval and release procedures. Third, the DBMSs should provide software tools that allow users to develop database management applications from scratch (Rouse, 1987).

A manufacturing engineering database can be logically and/or physically divided into several sub-databases. The types of databases can be either a relational model, a hierarchical model, or a network model. In many cases, a better idea is using a hybrid model, which combines relational and hierarchical or network models. A manufacturing engineering database which uses a hybrid approach may consist of the following four general sub-databases:

A design database which consists of several sub-databases such as geometrical modeling, bill of materials, group technology coding, etc. Documentation data can also be frequently accessed to its respective sub-database by design engineers (properties of materials, standards, handbooks, etc.). Test or quality measurement data can also be stored (tolerances, measures, surfaces roughness, etc.).

A process planning database which contains several sub-databases, such as raw materials, machine tools, tooling, and process plan (or route sheets). Each sub-database contains many different tables.

A manufacturing database which specifies how to manufacture a product or part, and is composed of a manufacturing database concerned with the NC programs, machine tools layout, tools, transfer devices, and material handling equipment. In addition, a manufacturing database includes a machinability database (the library) to store standard or constant technological data about work materials, tooling, tool geometry, and speed and feed rate tables.

A production management database which contains administrative sub-databases about the company products, personnel, customers, suppliers, status of inventories, equipment and processes, manufacturing orders, master schedule, and workshop loads.

In all of the above, each sub-database can be decomposed into further sub-databases or tables. In the following section, while we discuss the application of databases in process planning, we will introduce the hybrid approach database in detail.

3.4.3   Application of databases in process planning

Process planning is one of the major areas of application of database systems in manufacturing organizations. According to the general work steps of process planning that we have discussed in the previous chapter, databases for process planning can be built in many different ways. Whatever configuration the database is built into, some essential information must be provided, such as geometrical modeling, machine tools specification, raw material situation, tooling, and fixturing. All of these are necessary input information for process planning. When a database is built for process planning, the hybrid approach may be a better choice in which the general database and sub-databases are built in the form of a network model, and then each sub-database can contain further sub-databases or tables which are referred to the relational model.

Figure 3.8 illustrates a particular process planning database which uses a hybrid approach. There are generally five sub-databases. They are a raw materials database, machine database, tooling database, fixturing database, and cutting parameters database. Figure 3.8 indicates that the links between these databases are based on a network structure. It also indicates that some databases are related to other databases. For instance, the tooling database is the child node of both the raw materials database and the machine database. This means that the data for tools depends on both the data of materials to be cut and the machine tools to be employed. In addition, while cutting parameters information is based on the tool selected, specifications of tools are stored in the form of tables, which are relational models. To

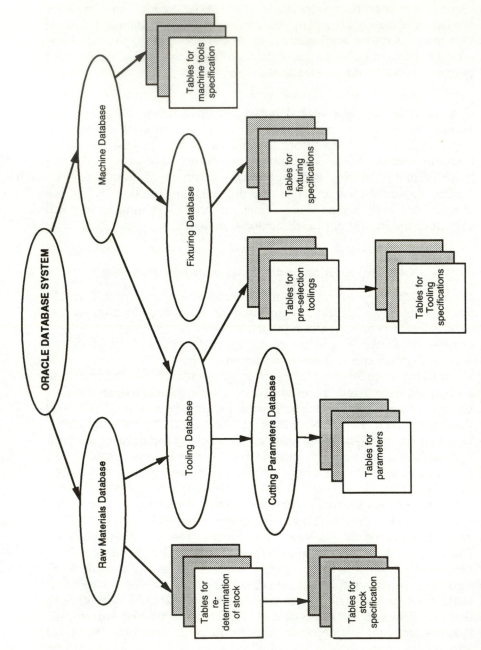

Figure 3.8 An example of a database for CAPP systems.

machi	machi ne name	machitype	spnbo	spnspeed	motor power	cost/m
TU01	BOEHRINGE VDF 400C	CNC LATHE	130.	14-2240	60KW	300.
TU02	BOXFORD 300-IS	CNC LATHE	35.	27-3000	1.5KW	180.
TU03	COLCHESTER CNC 200L	CNC LATHE	54.	20-2750	5KW	250.
TU04	CORTINI H105	CNC LATHE	20.	150-3000	1HP	205.
TU05	DYNA MYTE 3000	CNC LATHE	16.	0-4000	3/4 HP Servo motor	215.
TU06	MORI SEIKI CNC SL-5	CNC LATHE	77.	30-3000	DC22/30KW	270.
TU07	NAKAMURA SUPER TURN 2B	CNC LATHE	80.	10-3500	AC 11/15KW	280.
TU08	OKUMA LB15-1SC 500	CNC LATHE	56.	75-4200	11KW	290.
TU09	PARTNER L-100H	CNC LATHE	32.	80-4000	AC 5.5 KW	210.
TU10	YANAZAKI QUICK TURN 20	CNC LATHE	80.	13-3000	25 HP/18.7KW	270.

* *

machi	machine name	machdife	tolp.	spn. speed	motor power
ML01	CORTINI L300	CNC VERTICAL MINI MACHINING CENTER	6	150-3200	0.3KW (1.5HP)
ML02	FANUC TAPE CENTER-MODEL D	CNC MINI MACHINING CENTER	10	35-4500	7.5KW
ML03	HERMLE UWF 700 CNC	CNC MILLING AND DRILLING MACHINE	1	40-2000	2.2KW
ML04	MAZAK VQC-20/40	CNC DOUBLE-COLUMN MACHINING CENTER	20	28-4000	AC7.5HP/18KVA
ML05	MISUI SEIKI HR3B	CNC HORIZONTAL-SPINDEL MACHINING CENTER	30	45-4500	11/15 KW.AC
ML06	OKUMA MC-4VA	CNC VERTICAL MACHINING CENTER	20	50-4000	VAC3.7/5.5KW
ML07	PARTNER M-300	CNC VERTICAL MACHINING CENTER	20	100-6000	AC 3.7/2.2KW

Figure 3.9 An example of a machine database for CAPP systems.

get an explicit picture of this idea, three tables are provided here, which are directly printed out from the sub-databases.

The machine data contains specifications of all the machines that are used in the manufacture of a product. This data is categorized according to the type of machines, such as lathes, milling machines, drilling machines, etc. and is arranged in the form of a table. Each table contains the specifications of each machine in its columns. Figure 3.9 shows an example of machine tools data for CAPP systems.

The raw material data contains specifications of raw material, such as shape, size, etc., with a code number for each size of a particular shaped stock. The data is stored in different tables for different raw materials, such as steel, aluminum, plastic, etc. An example of a raw materials database is shown in Figure 3.10.

STOCK INVOICE   DATE: 08/02/89

• • • • • • • • • • • • • • • • •

Institute of Manufacturing Engineering
Technical University of Denmark
DK-2900 Lyngby, Denmark

— — — — — — — — — — — — — — — — — — — —
This STOCK INVOICE is used for the RWA MATERIALS DATABASE
of the expert process planning system XPLAN-R
— — — — — — — — — — — — — — — — — — — —

Property of stock: Medium carbon steel 45#

Codes	Feature	Dia.(mm)	Length(m)	Cost/m(DKK)	Stock state	Location
HXM45040	Hexagonal bar	40.	10.	285.	Available	MC.04.8
HXM45045	Hexagonal bar	45.	10.	285.	-0-	MC.04.9
SQM45020	Squared section	20.	5.	250.	Available	MC.05.3
SQM45025	Squared section	25.	5.	280.	-0-	MC.05.4
SQM45030	Squared section	30.	5.	285.	Available	MC.05.5
SQM45035	Squared section	35.	-0-	-0-	-0-	-0-
SQM45040	Squared section	40.	5.	320.	-0-	MC.05.6
SQM45045	Squared section	45.	-0-	-0-	-0-	-0-
BLM45020	Black bar	20.	8.	100.	Available	MC.01.4
BLM45025	Black bar	25.	8.	100.	Available	MC.01.5
BLM45030	Black bar	30.	8.5	110.	Available	MC.01.6
BLM45035	Black bar	35.	9.	110.	Available	MC.01.7
BLM45040	Black bar	40.	10.	120.	Available	MC.01.8
BLM45045	Bright bar	45.	10.	120.	Available	MC.01.9
BRM45020	Bright bar	20.	10.	150.	Available	MC.02.4
BRM45025	Bright bar	25.	10.	160.	Avialable	MC.02.5
BRM45030	Bright bar	30.	10.	165.	Available	MC.02.6
BRM45035	Bright bar	35.	10.	170.	Available	MC.02.7
BRM45040	Bright bar	40.	10.	175.	Available	MC.02.8
BRM45045	Bright bar	45.	10.	180.	-0-	MC.02.9
HXM45020	Hexagonal bar	20	8.	250.	Available	MC.03.4
HXM45025	Hexagonal bar	25.	8.	255.	-0-	MC.03.5
HXM45030	Hexagonal bar	30.	8.	265.	Available	MC.03.6
HXM45035	Hexagonal bar	35.	10.	275.	-0-	MC.03.7

Figure 3.10   An example of a raw material database for CAPP systems.

The tooling data contains the specifications of the tools permissible for certain cutting conditions, costs of the tools/minute, and other necessary information. An example of a tooling database in shown in Figure 3.11 (Zhang, 1989).

The detailed discussion about these sub-databases can be found in Section 11.6. Many efforts have been devoted and many suggestions have been raised towards an integrated manufacturing environment.

In general, a common database is required to store and retrieve the

Mostly often used raw materials

	Alloy steel	Alluminum alloy	Carbon steel	Composite/ Plastic	Copper/ Bronze	Gray cast iron	Stainless steel	. . . . . .
CF92								
FL11								
FS12								
FT52								
RS11			TC90CS			TC90GC		
. . . .								
XX00	XX00XX	XX00XX	XX00XX	XX00XX	XX00XX	XX00XX	XX00XX	. . . .

CODES I XX00: Four-alphanumeric-code to indicate the cutting subjects

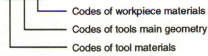

            Codes of cutting dimensions group

         Codes of cutting surface shape

      Codes of process activities

CODES II XX00XX: Six-alphanumeric-code to indicate the tool conditions

            Codes of workpiece materials

         Codes of tools main geometry

      Codes of tool materials

Figure 3.11 An example of a tooling database for CAPP systems.

common data from different production departments. For instance, some common databases can be used to store the data of geometrical models from a CAD system and can be accessed by a finite element analysis system. Since some software and hardware barriers exist currently, such a common database can only be shared within a stand-alone system, for example, geometrical model data, drafting data, and finite element analysis data within a CAD system; raw material data, machining data, and operation information data within a CAPP system. In order to reach the final integrated production, interfacing of these stand-alone common databases has to be considered. However, there are great difficulties in interfacing all of these separate activities. In order to realize the final integrated production, an ideal approach is to integrate all of the information involved in producing a product into a Single Data Base (SDB), instead of interfacing all of the stand-alone databases. The integrated SDB may include all the data from design, analysis, drafting, process planning, and NC tool paths, as well as the project planning, bill of materials, production scheduling information, post-process planning information, etc. In spite of great efforts devoted to improving the application of SDB, such an SDB system does not yet exist. Even the implementation technology is still in the early stages of development. Thus, based on the proposed principles, this type of SDB system may be considered a future target.

3.5 KNOWLEDGE BASED EXPERT SYSTEMS

Artificial Intelligence (AI) techniques have been recognized to have great potential for the development of Automated Process Planning (APP) systems. As a matter of fact, Knowledge Based Expert Systems (KBES) or simply Expert Systems (ES), as a branch of AI application, have been employed in many different areas. Many Expert Process Planning Systems (EPPS) have been developed throughout the world (Alting and Zhang, 1989; Davies, 1986; Davies *et al.*, 1986; Chang, 1989). In this section, a brief historical background of the development of expert systems is reviewed first. The fundamentals of knowledge based expert systems are then discussed in detail. This discussion includes the architecture of knowledge based systems, the approaches to knowledge acquisition and representation, the inference engine, the user interface scheme, etc. Finally, the application of expert systems in process planning is addressed.

3.5.1 Introduction

In 1950, Alan M. Turing posed the question 'Can machines think?' (Turing, 1950). For this and other contributions, Turing is generally recognized as the father of artificial intelligence. The term Artificial Intelligence (AI) was first proposed by John McCarthy at MIT at the end of the 1950s. He was, at that time, working with the development of the LISP-System that laid the foundation for this new research area (Sohlenius and Kjellberg, 1986). In almost the same period, Allen Newall and Cliff Shaw, who were employed by Rand Corporation, and Herbert Simon, who worked for Carnegie Mellon University (CMU), were working on the Logic Theory Machine (LTM) and their chess playing program, as well as the General Problem Solver (GPS) (Klahr and Waterman, 1986). This exploration was also defined as AI-related research during the early stage of AI development. Still, there is not a unified, comprehensive definition of AI. Basically, the term AI seems to be a kind of mechanism which is created by human beings to imitate some human mental activities. However, before the computer emerged, the idea of AI was only a scientific illusion. We might consider AI as an experimental branch of computer science which relates one complex natural phenomenon – human cognitive ability – with an artificial analogy – computer programs. AI studies the fundamental theory, technology, and methodology by which computer systems can be used to imitate human intelligent activities. Overall, AI may also be considered as a new scientific area. It is an area independent from computer science, and it explores the basic mechanism of intelligence, so that the human thinking process and intelligent activities can be imitated, or even created, reproduced, and reorganized by means of a machine that does not have to be a computer system. According to the different functions of simulating/imitating human activities, the application of AI can generally be divided into several different areas as follows:

- Pattern recognition
- Scene analysis
- Natural language understanding
- Game playing
- Automatic theorem proving
- Automatic programming
- Knowledge based expert systems
- Natural language synthesis
- Intelligent robots

AI may be divided into even more areas. However, knowledge based expert systems, or simply expert systems, are the largest branch of AI applications.

The first expert system, which is considered as the development of a clear set of general principles for designing an expert system, is probably DENDRAL (Hayes-roth *et al.*, 1983). The DENDRAL systems, consisting of Heuristic-DENDRAL and Meta-DENDRAL, were developed by a large research group which was established at Stanford University in 1965 to work on the design of a system to help chemists infer the structure of chemical compounds from mass spectral data. The Heuristic-DENDRAL made successful use of explicit expert knowledge of mass spectrometry and demonstrated the potential of the young discipline of AI to undertake problem solving in complex, real-world domains. Another initial well-known successful expert system is MYCIN (Alty and Coombs, 1984). MYCIN was developed at Stanford University in 1976 to assist physicians with advice on the diagnosis and treatment of infectious diseases. Its goals concerned the identification of the offending organism and the appropriate treatment (Hayes-roth *et al.*, 1983; Alty and Coombs, 1984; Tou, 1984; Klahr and Waterman, 1986; Bonnet *et al.*, 1988).

In general, the earliest successful applications of expert systems were in the chemical inference and medical diagnostic realms. These applications stimulated experiments in some other areas such as mineral prospecting, computer configuration, symbolic mathematics, chess playing, electronics analyses, and also in the industrial and manufacturing engineering areas (Hayes-roth *et al.*, 1983).

In the last two decades, expert systems have been defined in many different pieces of literature (Tou, 1984; Alty, 1985; Kumara *et al.*, 1986; Granville, 1986; Schaffer, 1986; Soliman, 1987; Wadhani, 1987; Pham and Pham, 1988; Zhang and Alting, 1989). Generally, an expert system is a computer system which has expert rules, avoids blind search, performs well, reasons by manipulating symbols, grasps fundamental domain principles, uses these in producing explanations, and has complete weaker reasoning methods to fall back on when expert rules fail. Expert systems deal with difficult problems in a complex domain, take a problem description in lay terms, and convert it to an internal representation appropriate for processing with expert rules. These systems can reason about their own knowledge (or lack thereof), especially to reconstruct inference paths rationally for explanation and self-justification. An expert system might work on one of the following types of tasks: diagnosis, monitoring, analysis, interpretation, consultation, planning, design, instruction, explanation, learning, conceptualization, control, prediction, debugging, or repair.

3.5.2 Architecture and implementation techniques

The role of an expert system is to store knowledge and to process it in such a way that it reaches the solution to a problem, using a method which can be explained to the user of the system. An expert system draws upon objective and subjective knowledge and manipulates it in a way that emulates the logic and reasoning processes used in problem solving by the human expert. Through the science of knowledge engineering, invaluable human expertise can be passed on to the computer. This is a vastly complex task. The construction of a successful expert system depends upon the accurate transfer of knowledge together with valid operation rules to process the knowledge correctly. In order to carry out this task, an expert system generally consists of the following components, which are illustrated in Figure 3.12:

- A knowledge base
- An inference engine

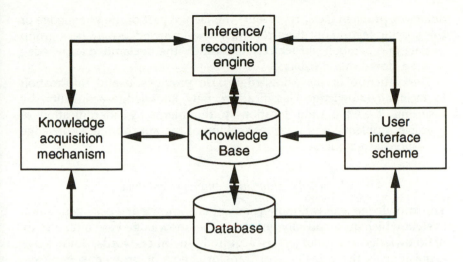

Figure 3.12 General architecture of an expert system.

- A user interface scheme
- A knowledge acquisition mechanism
- A database

Knowledge base

The knowledge base is the central component of an expert system. It contains generic knowledge for solving specific domain-related problems. According to the concepts of Knowledge Engineering (KE), the knowledge can naturally be subdivided into three broad categories. They are procedural knowledge, declarative knowledge, and control knowledge. Procedural knowledge and declarative knowledge are stored in the knowledge base (sometimes including a partial database for declarative knowledge). The control knowledge is used to construct the inference mechanism. The procedural knowledge, sometimes also called the knowledge sketch or production rules, represents the relational structure and problem orientated hierarchies of the knowledge stored or the production rules. In a natural format, the procedural knowledge may be represented in an associative hierarchical structure with semantic links. The most general concepts are placed at the top of the tree structure and the most specific items at the bottom of the hierarchy. Knowledge on various levels of specificity is distributed throughout the hierarchy. The knowledge hierarchy is linked to the declarative knowledge via semantic pointers.

The declarative knowledge, sometimes also called the knowledge

Impact of CIM

details or problem facts, represents the factual part of the knowledge or the specific features of the problems. The knowledge may be a group of data or a symbolic structure. Sometimes, the declarative knowledge may be stored in a database.

The structure of the knowledge base provides useful information to facilitate knowledge classification, new knowledge acquisition by inserting it at the right location in the hierarchy, identification of possible problems, and efficient seeking of the appropriate knowledge in response to a query.

Knowledge acquisition and representation

The knowledge acquisition mechanism is a knowledge base generation system which automatically generates the knowledge base after a body of knowledge on a subject is acquired. In other words, knowledge acquisition is the transfer and transformation of problem-solving expertise from some knowledge source to the system. Potential sources of knowledge include human experts, textbooks, databases, and one's own experience. Knowledge for an expert system can be acquired in several ways, all of which involve transferring the expertise needed for high performance problem solving in a domain from a source to the system. In general, the process of knowledge acquisition can be roughly characterized as problem implementation and testing. Knowledge acquisition is a bottleneck in the construction of an expert system. The knowledge engineer, who acts as a go-between to help an expert build a system, has far less knowledge of the domain than the expert. However, communication problems impede the process of transferring expertise into the system. A domain expert, who has the invaluable problem solving expertise and best understands the domain problem expertise, might not be good at computer programming implementation or operating disciplines. Therefore, the knowledge engineer and the expert must work together to develop an expert system. From this point of view, a person who is not only a skilled knowledge engineer but also an expert on the problem domain, will be the best candidate to join in the development work of expert systems.

In some cases, we might often hear of 'a frame-based expert system' or 'a rule-based expert system'. This is another way of classifying expert systems in terms of format for knowledge representation. The knowledge acquired should be represented in a computer implementation format. The representation of knowledge is a combination of data structures and interpretive procedures. There are many ways of representing knowledge, such as state space, logical representation schemes, semantic nets, frames, rule-based or hybrid approaches. In fact, the most popular approaches that have been addressed in the existing EPPS are the last four.

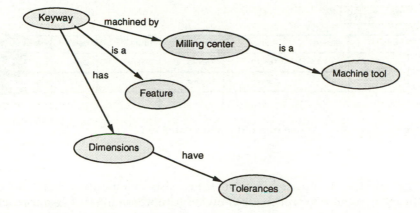

Figure 3.13   A simple semantic network data structure.

Semantic networks are based on the ancient and very simple idea that 'memory' is composed of associations between concepts. The basic concept of semantic networks is to describe the facts in terms of objects (nodes) which are linked by binary relations (arcs or labeled edges). The nodes represent the concept and the arcs represent the relationship between every two nodes. A simple semantic network's data structure is illustrated in Figure 3.13. Semantic networks are usually used to represent declarative knowledge. According to the information collected at the time this book was written, there is not yet any existing EPPS which employs the semantic networks approach.

The frame-based approach is sometimes also called pattern-directed. It is based on the principles of knowledge based pattern recognition (Tou, 1974). The frame-based approach organizes the knowledge that is to be stored in the knowledge base to form a prototype which is a record-like data structure. The knowledge is usually the declarative knowledge, which can be a class of objects, a form for encoding information on a stereotyped situation, a general concept, or a specific instance of any of these. Figure 3.14 shows a simple example of a frame data structure. The strength of the frame-based system lies in the fact that those items of declarative knowledge that are conventionally present in the description of an object or event are grouped together and may thus be accessed and processed as a unit. Many existing EPPSs are frame-based systems. One typical frame-based system is KAPPS (Iwata and Fukuda, 1987), which uses the frame to represent and acquire the know-how of process planning. Another frame-based system is SIPP (Nau and Chang, 1983). Both of these will be discussed in detail in a later section. Some database systems might also be considered as frame-based systems.

Name: Example

config.	Diameter	Process	Machines	Tools	Data	►Slots
Test 40	Real (MM)	Test 40	Test 40	Test 40	Data	►Types
Rotat.	30	Turning	Lathe A	TC90	11.09.89	►Fillers
.....	.....	.....	.....	.....	.....	.....

Figure 3.14   An example of a frame data structure.

The rule-based approach is sometimes also called rule-driven and is the most popular approach for knowledge representation. The approach is especially convenient to represent procedural knowledge. In a rule-based system, much of the knowledge in the knowledge base is represented as rules. These rules are expressed as conditional sentences relating statements of fact with one another. Therefore, in a rule-based expert system, the stored knowledge is often represented as a set of production rules in the form of an IF–THEN structure.

IF   ⟨antecedent 1⟩
          ⟨antecedent 2⟩
          .
          .
          .
          ⟨antecedent m⟩

THEN   ⟨consequent 1 with certainty C1⟩
          ⟨consequent 2 with certainty C2⟩
          .
          .
          .
          ⟨consequent n with certainty Cn⟩

The IF part is a set of conditions of a logical expression which must be satisfied to a certain pre-specified degree. The THEN part is a set of actions to be performed if the first part of the rule is satisfied. We may consider antecedents as patterns that can be matched against entries in the database and consequences as actions that can be taken or conclusions that can be inferred if all the antecedents match. These production rules reside in the knowledge base. A simple example of such an IF–THEN rule is as follows:

IF: It is an external diameter
        The diameter is less than 100 mm
THEN: A turning process is suggested
        A middle size lathe is required

Most of the existing EPPSs are based on the rule-based approach such as EXCAP (Davies and Darbyshire, 1984), GARI (Descotte and Latombe, 1981), TOM (Matsushima *et al.*, 1982), and XPLAN (Lenau and Alting, 1986; Alting *et al.*, 1988).

Hybrid approaches are a combination of frame-based and rule-based approaches. As mentioned before, the frame-based approach is especially convenient in representing the declarative knowledge, and the rule-based approach is especially convenient in representing the procedural knowledge. However, the knowledge (or expertise) of the process planning task includes both declarative and procedural knowledge. Some of the existing EPPSs employ both approaches to represent the corresponding knowledge. XPLAN-R is a typical example of such a system (Zhang and Alting, 1988a,b). XPLAN-R employs the DCLASS decision tree processor which performs IF–THEN rules to take the decision task of process planning, such as process selection, machine tools selection, operations, sequencing, etc., and then employs a relational database system, R:BASE System V, which stores and retrieves data in the form of tables that represent the specific data, such as machine tool specifications, stock situation, tooling, and fixturing description. Another example of the hybrid approach is the expert system OP for process selection (Lenau and Alting, 1988).

Inference engine

As mentioned previously, knowledge can be subdivided into three broad categories: declarative knowledge, procedural knowledge, and control knowledge. Declarative knowledge and procedural knowledge are stored in the knowledge base and represented in many ways. However, the control knowledge, which is the knowledge about a variety of processes, strategies, and structures used to coordinate the entire problem-solving process, is not stored in the knowledge base. Control knowledge forms another important component of an EPPS to generate alternative paths that lead to an inference. Such a component is called an inference engine or a reasoning mechanism. The inference engine is used to control the access to rules and facts. There are many different inference approaches, such as breadth-first, depth-first, forward and backward chaining, etc. Two main, popular strategies which are often used in inference are forward-chaining and backward-chaining.

Forward-chaining is also called forward-reasoning or data-driven inferring. It starts with a set of facts or given data and scans the rules of the condition part in the knowledge base. When the best suitable rule is found, the rule is applied, the knowledge base is updated, and the scanning resumes. This process continues until either a goal state is reached or no applicable rules are found. Some EPPSs, such as XPLAN, AMPS (Kanumury *et al.*, 1988), and CUTTECH (Barkocy and Zdeblick, 1984) are based on this reasoning strategy.

Backward-chaining is also called backward-reasoning or goal-driven inferring. It begins with the goal to be proven and tries to establish the facts needed to prove it by examining rules with the desired goal established as the consequence. If the condition part of a rule matches existing items in the knowledge base, then this rule is applied to solve the problem. If an unmatched antecedent is found, then arranging conditions to match that antecedent becomes a new sub-goal, and the same procedure is repeatedly applied. If there are no rules to establish the new sub-goal, then the system asks the user for more information and enters this into the knowledge base. Some EPPSs, such as EXCAP, TOM, SIPP and GARI, are based on this reasoning strategy.

User interface scheme

The user interface scheme is a collection of programs that work with the inference engine and the knowledge base to provide a convenient means of two-way communications with the user. The user interface gathers input data in one of two ways: the expert system may ask questions to which the user replies by typing in answers, or the user interface may operate by menus. These facilities offer multiple-choice questions, asking the user to select the correct choice from among several alternatives. This makes data entry fast and simple.

The user interface scheme also comes into play during the inferencing process. If the rule interpreter finds that it cannot reach a decision because it has insufficient information, then it notifies the user via the monitor. It may do this by picking up a portion of the rule being tested and reformatting it into a question. When the user enters the data asked for, the inference engine can continue its reasoning process.

Database

In addition to the above four necessary components (knowledge base, inference engine, knowledge acquisition mechanism, and user interface scheme), an expert system must have a database, sometimes called a global database or working memory, which is the portion of the computer's memory set aside for keeping track of inputs, intermediate conclusions, and outputs. The database contains the status of the program at any point in time. This makes it possible to stop the program at any point and interrogate the system to see what rules are being used and to trace the progress of the conclusion. It is like a scoreboard that keeps track of all the intermediate steps and conclusions as the program is being executed.

3.5.3 Applying expert systems in process planning

While there are many potential areas for the application of AI to manufacturing, such as monitoring, diagnosing, repairing, instructing, planning, and controlling, to name a few, the focus of this book is on the application of expert systems to manufacturing process planning. Not many years ago, only a few technical papers were published that discussed the application of expert systems to manufacturing process planning. However, up to 1988, more than 100 technical papers had been reported that discussed the application of expert systems to the manufacturing process planning function, and about 50 expert process planning systems (EPPS) had been developed throughout the world. In spite of this, existing EPPSs have not achieved the expected efficiency, but the initial limited success has attracted more and more interest from industrial researchers and engineers in developing EPPS. Manufacturing process planning has become one of the most active areas in expert systems application in manufacturing engineering. Some reasons for this situation are the following:

- With the rapid development of Computer Aided Design (CAD) and Computer Aided Manufacturing (CAM), CAPP is necessary to integrate CAD and CAM to reach the final Computer Integrated Manufacturing (CIM).
- The number of skilled industrial process planners is declining and to gather and save the expertise from those planners is an emergency task for almost every manufacturing company.
- The need for higher productivity has grown rapidly in modern factories in order to decrease cost, to reduce lead time, and to increase the quality of products. The traditional manual process planning approach cannot meet the requirements of competition (even for a sophisticated planner).
- The most important reason is that the task of process planning is a very complex one. Process planning uses not only working experience but also a variety of heuristic rules and a large number of logical decisions (the description of a process planner's task in detail can be found in section 2.5). Conventional computer algorithmic programs can do little with regard to the logical inference. Because of the partially heuristic and intuitive nature of the planning operation, it is not possible to generate adequate process plans solely by conventional algorithmic programs which manipulate data. Expert systems manipulate knowledge. That is the reason why the expert

system approach is so widely used in the experimental and practical stage of the process planning area.

To implement a computerized automated manufacturing process planning system, four basic models have to be considered. They are the workpiece model, the equipment model, the process model, and the process plan model (Spur, 1990). The shape is represented by the workpiece model. Accompanying the shape representation are the geometric and logical representations. The shape representation captures both the component topology and its geometry. The logical representation encompasses features, part and feature elements, and component assemblies. The graphical representation provides for the storage of the image of the component.

The workpiece model is composed of the three types of features – design, manufacturing, and geometric – as mentioned previously. These different features may be defined by different means. First, the features may be defined during the design process. Second, the simple features may be derived from simple algorithmic and analytical techniques. Finally, when features are more complex, they may be recognized as using knowledge-guided techniques. In expert systems for process planning, the knowledge for recognizing features is acquired from a human expert and coded into one of the knowledge structures described above. Again, the basic knowledge structures are rules, logic, frames, and semantic nets. Since many component features may be recognized from simple conditional logic, the rule-based paradigm is one of the most widely accepted. Frame-based methods, however, are rapidly gaining popularity and are the focus of a significant portion of current EPPS research. One of the initial requirements for a feature recognition system is that it can be independent of the geometrical representation of images by the CAD system. This integration of CAD and process planning has resulted in three streams of research (Spur and Specht, 1989). First, a conventional geometric modeling based CAD system can be linked to process planning systems via specially tailored interface software. Two basic methods have emerged to form this link: description languages and direct CAD system interfaces. A discussion of the advantages and disadvantages of each, as well as example systems, can be found in Gupta (1990).

Secondly, CAD systems may be integrated with knowledge-based systems. Spur *et al.* described this research as one in which solid CAD modeling systems are used for generating component geometry and providing geometric data, while the knowledge based environment is used to store related knowledge. The logical link between the two separate systems is partly formed by features (Spur *et al.*, 1989) which can be extracted using algorithmic programs. This approach is still primarily in the research stage.

The third CAD/CAPP system is one which develops CAD systems using knowledge based technologies and methods. As noted by Spur *et al.*, frame and object representations are used to store geometry as well as knowledge pertaining to constraints and to the relationships between geometric elements (Spur *et al.*, 1989; Spur *et al.*, 1990). Rules may also be used to determine factual data, to specify empirical relations, and to dictate imperatives. The development of an intelligent CAD system which directly supports process planning is only beginning to emerge.

Equipment models are the second of the four basic models that are utilized by CAPP systems. Equipment models are developed to relate the functionality and constraints of available machines to the process planning system. These equipment models must not only represent the workstations on the manufacturing floor but also represent machining tools, fixtures, and material handling and storage machines. Knowledge of both types of machines allows for the flow of processing to be coordinated with actual physical layouts and materials handling capabilities.

The logic for planning operation and process knowledge is represented in process models (Joshi, 1986; Spur, 1990). An operation is defined as the cutting of a given volume by a tool. This volume is defined as a manufacturing feature. A process is a composition of one or more operations. The process model must be able to relate each operation to the tool or tools which can perform the specified operation and to relate each operation to the fixture required to hold the component for processing. Also, associated with each operation and tools combination is an approximation of the time required to perform the operation. Processing time plays an essential role in planning through activity sequencing and scheduling. The final model required of CAPP systems is the process plan model. As Spur *et al.* noted, the logic of planning must represent the different levels of planning knowledge (Spur, 1990). Spur *et al.* also noted that the ultimate process planning goal may change, that the status of the manufacturing floor needs to be assessed, and that alternative process plans must be developed to achieve a production flexibility. Evaluating alternative process plans requires sophisticated simulation systems. As a matter of fact, many well known examples of expert process planning systems have been developed throughout the world. Some of these systems will be discussed in Chapter 9.

REFERENCES

Alting, L. (1986) Integration of engineering functions/disciplines in CIM. *Annals of the CIRP*, **35**(1).

Alting, L., Christensen, S.C. and Pedersen, M.A. (1987) An Integrated Miniature Laboratory for Research & Education. Isata, 17th International Symposium on Automotive Technology & Automation, Munich, West Germany, 25–30 October.

Alting, L. and Zhang, H.C. (1989) Computer Aided Process Planning: The state-of-the-art survey. *International Journal of Production Research*, 27(4), pp. 553–85.

Alting, L., Zhang, H.C. and Lenau, T. (1988) XPLAN – An Expert Process Planning System and its Further Development. Proceedings of 27th International MATADOR Conference, UMIST, UK, 20–21 April.

Alty, J.L. (1985) Use of expert systems. *Computer-Aided Engineering Journal*, February.

Alty, J.L. and Coombs, M.J. (1984) *Expert Systems Concepts and Examples*. NCC Publications, England, p. 209.

Barkocy, B.E. and Zdeblick, W.J. (1984) Knowledge-Based System for Machining Operation Planning. AUTOFACT'6, Anaheim California, October 1–4.

Bilberg, A. (1989) Simulation as a tool for developing FMS control programs. Ph.D. thesis, Institute of Manufacturing Engineering, Technical University of Denmark, Publication No. PI.89.02-A/AP.89.02.

Bjorke, O. (1984) Geometric modeling as an integral part of computer aided design and manufacturing systems. *Robotics & Computer-Integrated Manufacturing*, 1(2), pp. 161–5.

Bonnet, A., Haton, J.-P. and Truong-Ngoc, J.-M. (1988) *Expert Systems Principles and Practice*, Prentice Hall International Ltd, UK, p. 227.

Chang, T.-C. (1990) *Expert Process Planning for Manufacturing*, Addison-Wesley, p. 283.

Chang, T.-C., Anderson, D.C. and Mitchell, O.R. (1988) QTC – An Integrated Design/Manufacturing/Inspection System for Prismatic Parts. Proceedings of the 1988 ASME Computers in Engineering Conference, San Francisco, August 1–4.

Codd, E.F. (1970) A relational model of data for large shared data banks. *CACM*, 13(6), June.

Date, C.J. (1986) *An Introduction to Database Systems*, fourth edition, Addison-Wesley Publishing Company, USA, p. 639.

Davies, B.J. (1986) Application of expert systems in process planning. *Annals of the CIRP*, 35(2).

Davies, B.J. and Darbyshire, I.L. (1984) The use of expert systems in process planning. *Annals of the CIRP*, 33(1).

Davies, B.J., Darbyshire, I.L. and Wright, A.J. (1986) Expert Systems in Process Planning. Proceedings of the 7th International Conference on the Computer as a Design Tool, London, September 2–5.

Descotte, Y. and Latombe, J.C. (1981) GARI: A Problem Solver that Plans How to Machine Mechanical Parts. Proceedings of 7th International Joint Conference on Artificial Intelligence, Vancouver, Canada, August.

Draffan, I.W. and Poole, F. (1980) *Distributed Databases*, Cambridge University Press.

Granville, C. (1986) The Impact of Applying Artificial Intelligence with Group Technology. AUTOFACT'86, Detroit, Michigan, November 12–14.

Groover, M.P. (1987) *Automation, Production Systems, and Computer Integrated Manufacturing*, Prentice Hall, New Jersey.

Gupta, T. (1990) An expert system approach to process planning: current development and its future. *Computers in Industrial Engineering*, 18(1), pp. 69–80.

Hayes-roth, F., Waterman, D.A. and Lenat, D.B. (1983) *Building Expert Systems*, Addison-Wesley Publishing Company Inc., USA, p. 444.

Henderson, M.R. (1984) Extraction of feature information from three dimensional CAD data. Ph.D. thesis, Purdue University, West Lafayette, USA.

Iwata, K. and Fukuda, Y. (1987) KAPPS: Know-How and Knowledge Assisted Production Planning System in the Machining Shop. Proceedings of 19th CIRP International Seminar on Manufacturing Systems, Penn. State University, June 1–2.

Jakubowski, R. (1982) Syntactic characterization of machine part shapes. *Cybernetics and Systems Int. J.*, **13**.

Joshi, S. and Chang, T.-C. (1988) Graph-based heuristic for recognition of machined features from a 3D solid model. *Computer-Aided Design*, **20**(2).

Joshi, S., Chang, T.C. and Liu, C.R. (1986) Process planning formalization in an AI framework. *International Journal of AI in Engineering*, **1**(1), pp. 45–53.

Kanumury, M., Shah, J. and Chang, T.C. (1988) An Automatic Process Planning System for a Quick Turnaround Cell – An Integrated CAD and CAM System. USA–Japan Symposium on Flexible Automation, July 18.

Klahr, P. and Waterman, D.A. (1986) *Expert Systems Techniques, Tools, and Applications*. Addison-Wesley Publishing Company, USA, p. 441.

Kochan, D. (1986) CAM Development in Computer-Integrated Manufacturing, Springer-Verlag, Berlin, p. 289.

Kochhar, A.K. (1979) *Development of Computer-Based Production Systems*, Edward Arnold, London, p. 274.

Kumara, S.R.T., Joshi, S., Kashyap, R.L., Modie, C.L. and Chang, T.C. (1986) Expert systems in industrial engineering. *International Journal of Production Research*, **24**(5), pp. 1107–25.

Kung, H. (1984) An investigation into development of process plan from solid geometric modeling representation. Ph.D. thesis, Oklahoma State University, USA.

Kyprianou, L.K. (1983) Shape classification in computer aided design. Ph.D. thesis, Christ College, University of Cambridge, UK.

Laing, M., Dutta, S.P. and Abdou, G. (1989) A New Approach to Material Handling Equipment Selection in a Manufacturing Environment. International Industrial Engineering Conference, May 14–17.

Lenau, T. and Alting, L. (1986) XPLAN – An Expert Process Planning System. Proceedings of the 2nd International Expert Systems Conference, London, UK, 30 September–2 October.

Lenau, T. and Alting, L. (1988) Artificial Intelligence for Process Selection. CIRP International Seminar on Manufacturing Technology, Singapore, 18–19 August.

Matsushima, K., Okada, N. and Sata, T. (1982) The integration of CAD and CAM by application of artificial-intelligence techniques. *Annals of the CIRP*, **31**(1).

Monostori, L., Zsoldos, L. and Bartal, L. (1989) An Approach to Knowledge Based, Comprehensive Monitoring and Diagnostics of Manufacturing Cells. 21st International Seminar on Manufacturing Systems, June 5–6, Stockholm, Sweden.

Nau, D.S and Chang, T.C. (1983) Prospects for process selection using artificial intelligence. *Computers in Industry*.

Pham, D.T. and Pham, T.N. (1988) Expert systems in mechanical and manufacturing engineering. *The International Journal of Advanced Manufacturing Technology*, **3**(3), July.

Ranky, P.G. (1986) *Computer Integrated Manufacturing*, Prentice Hall, p. 509.

Rouse, N.E. (1987) Managing engineering databases. *Machine Design*, September 10, pp. 108–9.

Schaffer, G.H., ed. (1986) Artificial intelligence. *American Machinist & Automated Manufacturing*, August.

Sohlenius, G. and Kjellberg, T. (1986) Artificial intelligence and its potential use in the manufacturing system. *Annals of the CIRP*, **35**(2).

Soliman, J.T. (1987) Expert Systems in Manufacturing. Proceedings of the 17th ISATA Conference, Keynote Speeches, Munich, West Germany.

Spur, G., Krause, F.-L. and Kempf, M. (1990) Advanced Software Tools for the Development of CAPP Systems. Proceedings of 22nd CIRP International Seminar on Manufacturing Systems, Enschede, The Netherlands, June 11–12.

Spur, G., Krause, F.-L. and Major, F.W. (1989) Advanced Software Technologies for CAPP. CIRP International Workshop on CAPP, Hanover, Germany, September 21–22.

Spur, G. and Specht, D. (1989) Knowledge-Based Diagnosis in Manufacturing Systems. 21st International Seminar on Manufacturing Systems, June 5–6, Stockholm, Sweden.

Tou, J.T. (1974) *Pattern Recognition Principles*, Addison-Wesley Publishing Company, Reading, Massachusetts, USA.

Tou, J.T. (1984) Design of expert system for integrated production automation, *Journal of Manufacturing Systems*, **4**(2).

Turing, A.M. (1950) Computing machinery and intelligence. *Mind*.

Wadhani, R. (1987) Artificial Intelligence in the Factory of Today. Proceedings of the 17th ISATA Conference, Keynote Speeches, Munich, West Germany.

Zhang, H.C. (1989) Computerized manufacturing process planning for rotational components. Ph.D. dissertation, Institute of Manufacturing Engineering, Technical University of Denmark, Publication No. PI.89.14-A/AP.89.27.

Zhang, H.C. and Alting, L. (1988a) Introduction to an intelligent process planning system for rotational parts. *ASME-WAM 1988 PED* **31**, Chicago, Illinois, November 27–December 2.

Zhang, H.C and Alting, T. (1988b) XPLAN-R An Expert Process Planning System for Rotational Components. Proceedings of IIE 1988 Integrated Systems Conference, St Louis, Missouri, USA, October 30–November 2.

Zhang, H.C and Alting, L. (1989) Expert Process Planning Systems: The State-of-the-Art. Proceedings of 21st CIRP International Seminar on Manufacturing Systems, Stockholm, Sweden, June 5–6.

CHAPTER 4

Group technology in process planning

4.1 INTRODUCTION

Group Technology (GT) is one of the most important technologies which impact the task of process planning. Application of GT in process planning has changed process planning dramatically, provided significant convenience, and brought about considerable economic benefits for process planning. As mentioned before, mass production or flow-type manufacturing offers the greatest economies of scale. If we are involved in a business that produces products that are virtually identical in mass quantities, we have the opportunity to employ special-purpose machines optimized to produce the product. However, mass production accounts for only 30% of all manufacturing. The remaining 70% of manufacturing is accomplished in job shops or batch size production which is done on relatively general-purpose machine tools. Also in job shop or batch size production, the cost of optimization is usually not economically sound, and many unproductive compromises must be made. Therefore, much thought has been given to finding ways to make job shop and batch size production techniques approach those of flow manufacturing. Among all these attempts, group technology (GT) is the dominant one; it has gained total acceptance and been widely used all over the world.

4.2 HISTORICAL BACKGROUND AND DEFINITION

Some people believe that John Mitrofanov, a Russian engineer, introduced the concept of GT, while others tend to believe that it was not Mitrofanov but his teacher Sokolowski from Leningrad. Tuffentsammer and Arndt (1983) wrote that Sokolowski's doctoral dissertation was not translated into English and German as was the book written by Mitrofanov, and this is the reason why Mitrofanov is often given credit for the development of GT instead of Sokolowski. Tuffentsammer and Arndt also wrote that 'It was already Sokolowski's basic premise to take the use of the same production machinery, to be considered for the machining of similar parts, as the decisive measure of the relevant similarity features of the parts which were to be machined on the same production equipment, is possible without resetting.' (Tuffentsammer and Arndt, 1983).

Whichever story you choose to believe is purely personal, the fact is that GT was introduced approximately 50 years ago in Russia. Unfortunately, this idea was not widely adopted until the beginning of the 1970s. Since computer aided technologies and flexible manufacturing systems (FMS) were introduced into manufacturing industry, GT has become recognized and is widely utilized.

There have been many definitions of group technology. Group technology has been described as a concept, organizational principle, discipline, philosophy, and methodology. In fact, GT is all of these. GT is defined in this book as 'the philosophy of studying a large population of apparently different items and then dividing them into groups of items having the same or similar characteristics'.

4.3   CLASSIFICATION AND CODING

The typical utilization of GT is in the discrete parts manufacturing enterprise where the parts are grouped into part families by means of coding and classification. A part family is a collection of parts which are similar either because the geometrical shapes and sizes of the parts are similar or because the processing steps required in their manufacture are similar. The parts within a family are different, but their similarities are close enough to merit their identification as members of the part family. Figure 4.1 illustrates some rotational part families.

In general, five methods are used to form part families:

- **Manual/visual search:** Ocular grouping, better known as 'eye balling', is a popular way of grouping parts, but it has many limitations. This method will become very awkward and inefficient when the number of parts is increased.
- **Nomenclatures/functions:** Names are a useful way of identifying people. This idea is borrowed for classification and coding. In some sense, this is an efficient way, but there are several limitations when it comes to identifying parts. The problem is similar to that of the manual/visual method; when there is a large number of parts, this method will cause some confusion for retrieving part families with similar or close names.
- **Production flow analysis:** All of the parts that are to be produced by a given group of machine tools or flexible manufacturing cells should be grouped together. This method provides significant convenience for manufacturing engineers but does not meet designing

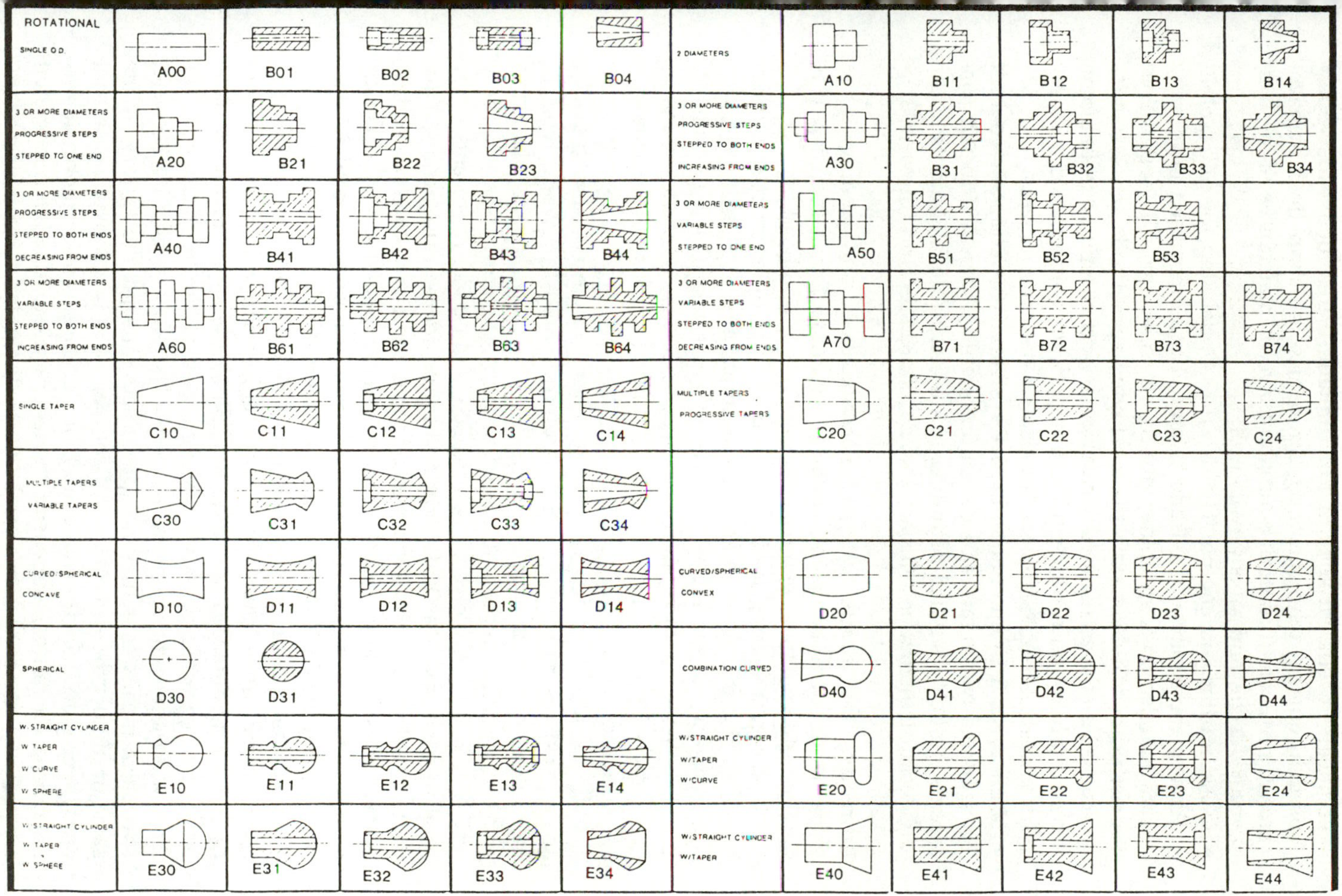

Figure 4.1 Examples of rotational part families.

engineers' needs. In spite of this limitation, this method is widely used in current manufacturing companies for shop floor management, production control, equipment layout, material handling conveyor design, and inventory control.

- **Classification and coding:** This is an essential step for full exploitation of the benefits of GT and is more commonly used in forming part families and machine groups or cells for GT application. Classification and coding is the most popular method used in today's companies. Especially when combined with computer technology, classification and coding provides significant convenience not only for manufacturing engineers but also for design engineers.
- **Mathematical programming/expert systems:** This is a new concept for generative coding and computer-oriented schemes. Expert systems are being studied according to the recent developments in the CIM implementation (Kusiak, 1987, 1988; Kusiak and Chow, 1987).

Currently, the fourth category, classification and coding, is a sophisticated approach being applied in manufacturing industry. It is necessary to give more emphasis to classification and coding.

Classification is used to divide the parts into different part families based on the existence or absence of similarities among these parts. Coding is used to entitle and identify the part similarities with symbols. Usually part similarities are of two types: design attributes (such as geometric shape and size) and manufacturing attributes (the sequence of processing steps required to make the part). A coding system is a series of alphanumeric characters. There are three basic types of coding systems namely monocode, polycode, and the hybrid type. In monocode, each code number is qualified by the preceding character, and can therefore represent a large amount of information with very few code positions. Polycode, also called chain code, represents every single digit despite the precedent and is considered compact and easy to construct. The hybrid type is a combination of the other two coding systems and encompasses the merits of both systems. Some famous classification and coding systems will be discussed in detail in later sections. Figure 4.2 illustrates an example of classification codes.

When components are classified and coded, the classification and coding can be done in one of two ways, either from a design standpoint or from a manufacturing standpoint. Since the scope of this book is focused on process planning, we will look at the application of GT from the manufacturing point of view. Therefore, similarity between two parts means that they are processed through the factory in the same or almost the same way. Of course, parts that look alike are not always produced in the same way (it depends on variations in raw materials, tolerances, dimensions, etc.), while parts that are routed through the same machines can be quite dissimilar in geometric form.

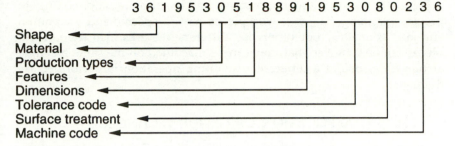

3 6 1 9 5 3 0 5 1 8 8 9 1 9 5 3 0 8 0 2 3 6

Shape
Material
Production types
Features
Dimensions
Tolerance code
Surface treatment
Machine code

Figure 4.2 An example of classification codes.

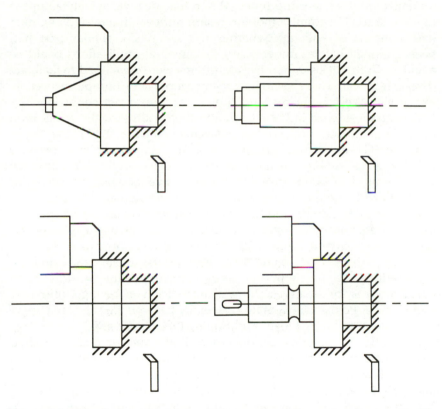

Figure 4.3 Identical machining operation performed on workpieces with different end shapes (Tuffentsammer and Arndt, 1983).

A classification and coding system is used to give each part a numeric or alphanumeric code describing the points of interest of each part. Usually, the classification and coding system that works best is the one that can describe a part from both the design and manufacturing point of view.

Figure 4.3 shows the importance of classification and coding to process planning. The two workpieces have a specific and a common end-face geometry, but otherwise different features. The two workpieces can be coded so that they can be machined on the same machine at one time, before a different machining operation is performed on them.

4.4 APPLICATION OF GROUP TECHNOLOGY

At the beginning, the realm of application of the GT concept was not as wide as the current situation. Mitrofanov first proposed using GT to minimize production setup times. When this idea was widely accepted in the world, GT became a tool for design retrieval. If a design engineer had a means of knowing whether the same or a similar part had been designed before, unnecessary designs and duplications could be avoided. In the 1950s, a number of systems were developed to facilitate design retrieval. These systems became popular in Europe, especially in the UK, West Germany, and the eastern European countries. The University of Aachen in West Germany then addressed the application of GT in manufacturing processes (Teicholz and Orr, 1987). The application of GT in manufacturing was a very important step because it led to a realization that the inexpensive mass production approach could be used as a substitute for the expensive job shop or batch size production. In addition to the work in West Germany, efforts were made in the UK, France, and the eastern European countries to optimize batch production by using GT concepts. Since batch size production is the most common form of production throughout the world, constituting perhaps 70% or more of the total manufacturing activities, these efforts led to significant improvements in the world's batch production. Since the mid 1970s, Allen, of Brigham Young University, and Ham, of Pennsylvania State University have pushed the idea of GT for American industry (Allen and Smith, 1980; Ham, 1982).

Hyer and Wemmerlov conducted a study and predicted in their paper MRP/GT: A Framework for Production Planning and Control of Cellular Manufacturing (Hyer and Wemmerlov, 1982) that between 50 and 70% of American industries would use GT by 1990. There has been no recent survey on the percentage of companies using GT, but the indications from the studies done thus far indicate that the prediction is correct, or perhaps even lower than the actual percentage of companies using GT.

The application of GT concepts is found in a number of activities including design, process planning, inventory ordering and control, and part manufacturing.

Many existing process planning systems are based on variants of GT. The DCLASS system, which is widely used for GT classification and coding, was developed by Brigham Young University (Allen, 1980), and MULTICLASS II and MULTICAPP II, which are two parts of a broad-spectrum GT system, were developed by the Organization for Industrial Research (OIR) (OIR Product News, 1987). In the meantime, GT has also been introduced into generative systems, and fairly good results have been established in the CIM concepts. For instance, an integrated system entitled PFDM (Part Family Design and Manufacturing) based on integration of CAD–CAPP–NC has been established at the Technical University of Denmark (Alting and Lenau, 1986). The application of GT now covers almost all of the realms of the manufacturing functions, namely:

- Product design
- Tooling and setups (process flow)
- Materials handling (equipment layout)
- Production and inventory control
- Process planning

Several differently oriented classification and coding systems have been used in real production environments (Halevi, 1980), including:

- Design-oriented classification and coding systems
- Production-oriented classification and coding systems
- Resource-oriented classification and coding systems

The approaches of resource-oriented classification and coding systems form the concept of Cellular Manufacturing (CM) or machining cells, which are fundamentals of FMS. Application of cellular manufacturing ideas significantly increases the productivity and flexibility for batch production. Recently, the fifth category – mathematical programming/ expert systems – is also reported to have made good progress. An integrated approach to GT part family database design based on AI techniques has been developed at Pennsylvania State University (Ham *et al.*, 1988). Another example is SAPT – an expert system based on a hybrid concept of GT reported by Milacic (Milacic *et al.*, 1987). Kusiak has reported the results of several research projects that used expert systems in GT (Kusiak, 1987, 1988, 1989, 1990). The application of GT leads to an Automated Process Planning System (APPS) which will be discussed in section 9.3.

4.5 SOME EXAMPLES OF CLASSIFICATION AND CODING SYSTEMS

4.5.1 Vuoso–Praha system

The Vuoso–Praha coding system uses a four-digit code to represent parts according to type, class, group, and material. The Vuoso–Praha

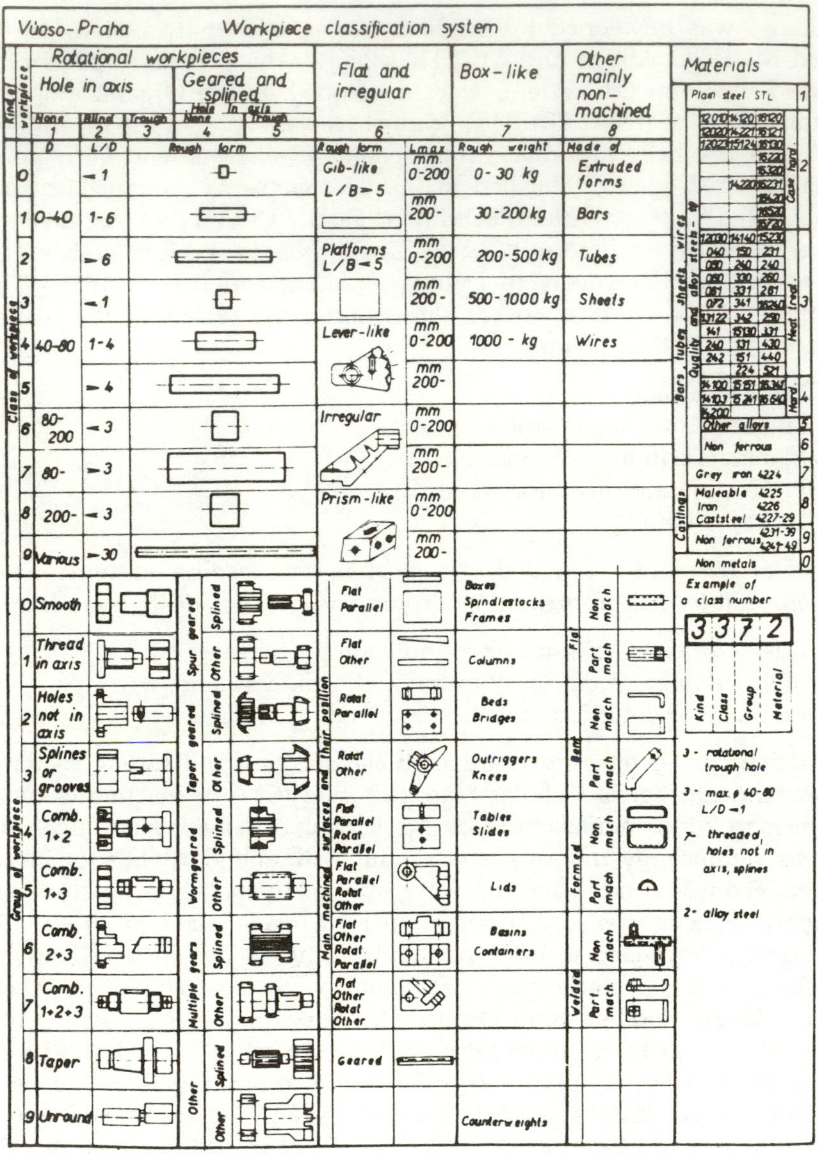

Vuoso–Praha   Workpiece classification system

	Rotational workpieces		Flat and irregular	Box–like	Other mainly non–machined	Materials
	Hole in axis / Geared and splined					Plain steel STL — 1

Figure 4.4 An example of the Vuoso–Praha coding system. *Source*: Group Technology, C.C. Gallagher and W.A. Knight (1973).

coding system is 'typically used for rough part classification so as to identify the type of department that would produce the part'. (Chang *et al.*, 1991).

An example of the Vuoso–Praha coding system is shown in Figure 4.4. Notice from the example how little information is obtained about the component after classification. This is not the case in coding systems such as KK-3 or even the Opitz system, but as stated previously, the Vuoso–Praha coding system is a 'rough part classification system'. Therefore, those companies needing more details about a coded component should look at coding systems such as KK-3, Opitz, MICLASS, etc.

4.5.2   KK-3 system

The KK-3 coding system was developed by the Japan Society for the Promotion of Machine Industry (JSPMI) and was first presented in 1976. The KK-3 system is basically a general-purpose classification and coding system that uses 21 digits to code predominantly metal-cutting and grinding components (Chang *et al.*, 1991).

A structure of the KK-3 system is shown in Figure 4.5. The KK-3 system is structured such that much more detail about a part can be represented in the coding. For example, KK-3 can classify 100 functional names for rotational and non-rotational components. An example of how the KK-3 coding system works is shown in Figure 4.6. Notice how 13 digits (#5 to #17) are used to classify shape detail and process type. Even more information can be given about a component using the KK-3 system compared to the CODE or Opitz system (Chang *et al.*, 1991).

4.5.3   MICLASS/MULTICLASS system

The MICLASS system is a chain-structured coding system which uses 12 digits to represent the component part. An additional 18 digits are available for user-specified information, such as lot size, part function, etc. The MICLASS system was originally developed by TNO of Holland and is currently maintained in the United States by the Organization for Industrial Research (OIR). MULTICLASS II is the latest version of MICLASS, which makes it possible to enter part descriptions quickly and easily in a tightly organized framework that effectively supports retrieval and analysis. MULTICLASS II can be used to retrieve information and to form part families based on common attributes. The user can retrieve parts by keywords, features, individual characteristics, part numbers, classification numbers, or other data. The system uses common terms tailored to individual company practices (OIR Group Technology, 1986).

Digit	Items	(Rotational Components)		
1	Parts	General classification		
2	name	Detail classification		
3	Materials	General classification		
4		Detail classification		
5	Chief	Length		
6	dimensions	Diameter		
7	Primary shapes and ratio of major dimensions			
8	Shape details and kinds of process	External surface	External surface and outer primary shape	
9			Concentric screw threaded parts	
10			Functional cut-off parts	
11			Extraordinary shaped parts	
12			Forming	
13			Cylindrical surface	
14		Internal surface	Internal primary shape	
15			Internal curved surface	
16			Internal flat surface and cylindrical surface	
17		End surface		
18		Non-concentric holes	Regular located holes	
19			Special holes	
20		Non-cutting process		
21	Accuracy			

Figure 4.5 Structure of KK-3 system.

The MICLASS system is so designed that it can be used to include both design and manufacturing information. The system has four major parts:

- **The classification number**, which makes it possible to classify parts by their engineering and manufacturing characteristics
- **The database**, which contains design and manufacturing information for the whole company
- **Retrieval programs**, which make it possible to retrieve drawings, route sheets, manufacturing instructions, etc.
- **The analysis programs**, which are used for design standardization, optimization of machine tool use, and manufacturing routines.

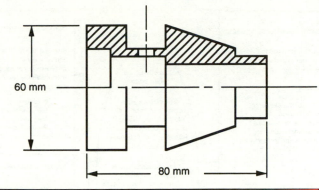

Code digit	Item	Component condition	Code
1	Name	Control valve	0
2		Others	9
3	Materials	Copper bar	7
4			5
5	Dimension length	80 mm	2
6	Dimension diameter	60mm	2
7	Primary shape and ratio of chief dimension	L/D 1.3	2
8	External surface	With functional tapered surface	3
9	Concentric screw	None	0
10	Functional cut-off	None	0
11	Extraordinary shaped	None	0
12	Forming	None	0
13	Cylindrical surface>3	None	0
14	Internal primary	Piercing hole with dia. Variation, no cut-off	2
15	Internal curved surface	None	0
16	Internal flat surface	None	0
17	End surface	Flat	0
18	Regularly located hole	Holes located on circumferential line	3
19	Special hole	None	0
20	Non-cutting process	None	0
21	Accuracy	Grinding process on external surface	4

Figure 4.6 Example of a KK-3 coding system.

Actually, MULTICLASS II is an integral part of a series of software packages for the application of GT which is named Multi-II Group Technology system, including MULTICLASS II, MULTIGROUP II, MULTITRIEVE II, MULTICAPP II, MULTICATS II, and MULTITRACK II. See Figure 4.7 for an illustration of the MICLASS system.

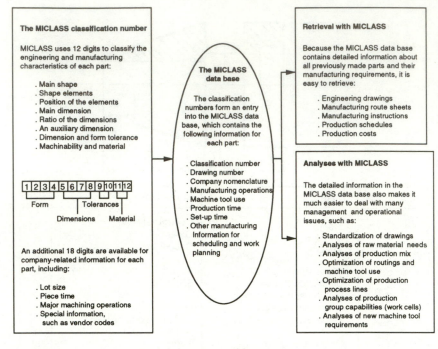

The MICLASS classification number

MICLASS uses 12 digits to classify the engineering and manufacturing characteristics of each part:

. Main shape
. Shape elements
. Position of the elements
. Main dimension
. Ratio of the dimensions
. An auxiliary dimension
. Dimension and form tolerance
. Machinability and material

| 1 | 2 | 3 | 4 | 5 | 6 | 7 | 8 | 9 | 10 | 11 | 12 |

Form   Tolerances

Dimensions   Material

An additional 18 digits are available for company-related information for each part, including:

. Lot size
. Piece time
. Major machining operations
. Special information, such as vendor codes

The MICLASS data base

The classification numbers form an entry into the MICLASS data base, which contains the following information for each part:

. Classification number
. Drawing number
. Company nomenclature
. Manufacturing operations
. Machine tool use
. Production time
. Set-up time
. Other manufacturing information for scheduling and work planning

Retrieval with MICLASS

Because the MICLASS data base contains detailed information about all previously made parts and their manufacturing requirements, it is easy to retrieve:

. Engineering drawings
. Manufacturing route sheets
. Manufacturing instructions
. Production schedules
. Production costs

Analyses with MICLASS

The detailed information in the MICLASS data base also makes it much easier to deal with many management and operational issues, such as:

. Standardization of drawings
. Analyses of raw material needs
. Analyses of production mix
. Optimization of routings and machine tool use
. Optimization of production process lines
. Analyses of production group capabilities (work cells)
. Analyses of new machine tool requirements

Figure 4.7 An illustration of the MICLASS system.

4.5.4 DCLASS system

DCLASS, which stands for Decision CLASSification Information System, was developed by D.K. Allen and P.R. Smith at Brigham Young University in 1980. It is a computerized method to classify, store, and retrieve information efficiently and rapidly, to do calculations, and to aid in decision-making. (Allen and Smith, 1980).

At the heart of the DCLASS system are Trees designed to show relationships and processes. There are five different Tree types with various branch (node) configurations. The five types of Trees are:

- **E-Tree** (mutually exclusive branching)
- **N-Tree** (non-mutually exclusive multiple path)
- **C-Tree** (combination of E and N Trees)
- **X-Tree** (computer mathematical expressions)
- **D-Tree** (decision tree If . . . Then)

The DCLASS system operates by allowing the user first to design a tree(s) for the part(s), apply the coding syntax to the tree(s), and then enter the program into the computer by way of a tree source file. Then the user runs the program to check for errors, and finally tries out the system (Allen and Smith, 1980). An example of how the process is

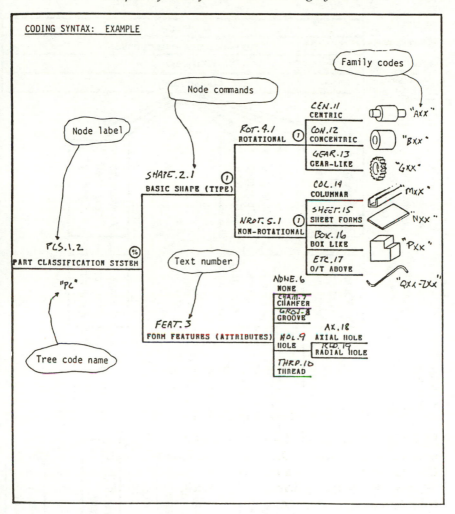

Figure 4.8   DCLASS flow chart.

done is shown in Figure 4.8. The DCLASS system is very easy to use, and someone with almost no computer knowledge can design the trees and enter the program into the computer. DCLASS is very beneficial to companies in that it increases productivity and reduces costs. Another benefit of the DCLASS system is that it is easier to work with than other hard-coded coding systems.

4.5.5   Opitz system

The Opitz coding system was developed in 1970 by H. Opitz of the Aachen Technical University in West Germany. The Opitz code is one of

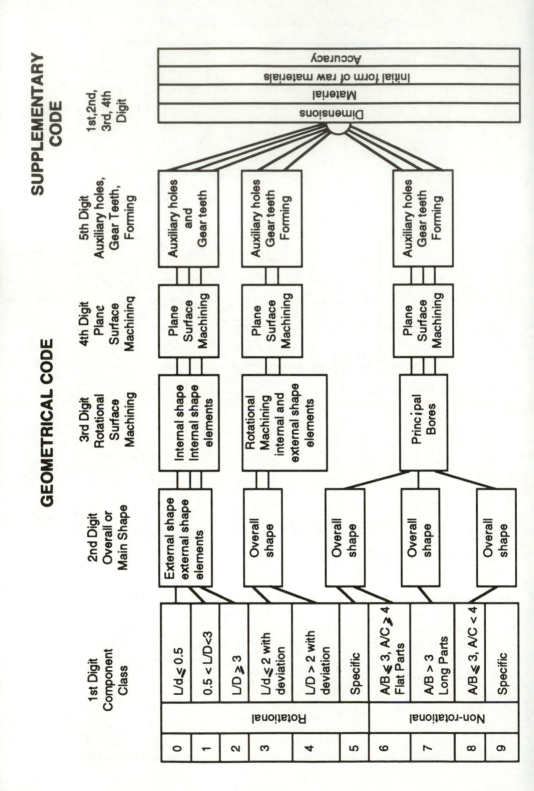

Digit 1

	Diameter d or edge length A	
	mm	inches
0	<20	<0.8
1	>20 ⩽50	>0.8 ⩽2.0
2	>50 ⩽100	>2.0 ⩽4.0
3	>100 ⩽160	>4.0 ⩽6.5
4	>160 ⩽250	>6.5 ⩽10
5	>250 ⩽400	>10 ⩽16
6	>400 ⩽600	>16 ⩽25
7	>600 ⩽1000	>25 ⩽40
8	>1000 ⩽2000	>40 ⩽80
9	>2000	>80

Digit 2

	Material
0	Cast iron
1	Modular graphitic cast iron and malleable cast iron
2	Mild steel <26.5 tonf/sq. in not heat treated
3	Hard steel ≥26.5 tonf/sq. in heat-treatable low-carbon and case-hardening steel, not heattreated
4	Steel 2 and 3 heat treated
5	Alloy steel (not heat treated)
6	Alloy steel heat treated
7	Nonferrous metal
8	Light alloy
9	Other materials

Digit 3

	Initial form
0	Round bar, black
1	Round bar, bright drawn
2	Bar: triangular, square, hexagonal, others
3	Tubing
4	Angle, U-, T-, and similar sections
5	Sheet
6	Plate and slabs
7	Cast or forged components
8	Welded assembly
9	Pre-machined components

Digit 4

	Diameter D or edge length A
0	No accuracy applied
1	2
2	3
3	4
4	5
5	2 and 3
6	2 and 4
7	2 and 5
8	3 and 4
9	2 + 3 + 4 + 5

Figure 4.10 Form code for rotational parts in the Opitz system.

	Digit 1	Digit 2	Digit 3	Digit 4	Digit 5
	Part class	External shape, external shape elements	Internal shape, internal shape elements	Plane-surface machining	Auxiliary holes and gear teeth
0	L/D ≤ 0.5	Smooth, no shape elements	No hole, no breakthrough	No surface machining	No auxiliary hole
1	0.5 < L/D < 3	No shape elements	No shape elements	Surface plane and/or curved in one direction, external	Axial, not on pitch circle diameter
2	L/D > 3	With screw thread	Thread	External plane surface related by graduation around a circle	Axial on pitch circle diameter
3		With functional groove	Functional groove	External groove and/or slot	Radial, not on pitch circle diameter
4		No shape elements	No shape elements	External spline (polygon)	Axial and/or radial and/or other direction
5		Thread	Thread	External plane surface and/or slot, external spline	Axial and/or radial on PCD and/or other direct.
6		Functional groove	Functional groove	Internal plane surface and/or slot	Spur gear teeth
7		Functional cone	Functional cone	Internal spline (polygon)	Bevel gear teeth
8		Operating thread	Operating thread	Internal and external polygon, groove, and/or slot	Other gear teeth
9		All others	All others	All others	All others

Rotational parts: 0, 1, 2, 3, 4, 5

Non-rotational parts: 6, 7, 8, 9

Digit 2 — Stepped to one end or smooth: 1, 2, 3; Stepped to both ends: 4, 5, 6

Digit 3 — Smooth or stepped to one end: 1, 2, 3; Stepped to both ends: 4, 5, 6

Digit 5 — No gear teeth: 0, 1, 2, 3, 4, 5; With gear teeth: 6, 7, 8, 9

Figure 4.10 Continued

the most popular and well known coding systems because it is concise and easy to use.

The Opitz code is a nine-digit coding system that uses a geometric code for the first five digits and a supplementary code for the last four digits. The code can be used to classify rotational or non-rotational parts. Rotational parts are classified by their length/diameter ratio and non-rotational parts are classified by their length/width and length/ height ratios.

The first five digits of the Opitz code represent: (1) part class, (2) external shape, external shape elements, (3) internal shape, internal shape elements, (4) plane surface machining, and (5) auxiliary holes and gear teeth. The last four digits represent (1) diameter d or edge length A, (2) material, (3) initial form, and (4) diameter D or edge length A (Accuracy). Figure 4.9 shows how the Opitz coding system is broken down into nine digits, while Figure 4.10 shows how the digits are selected according to the part class, external shape, etc. (Opitz, 1970).

The Opitz coding system is currently used by several companies as their CAPP coding systems because it is concise and easy to use.

Example of using Opitz system for shaft part, screw part, flange part

The three examples – shaft (ZT-SHA), screw (ZT-SCR), and flange (ZT-FLA) – are shown in Figures 4.11, 4.12 and 4.13, respectively. The Opitz coding system is used to code the three examples, and the codes are given along with a brief description of each part.

ZT-SHA is a shaft part made of mild steel bright bar. It is a single axis multi-diameter shaft with several external features. The external features have seven different diameters: one with two chamfers and gears, one with a 5 mm fillet, one with a keyway, and one with two chamfers. The Opitz code found for the shaft part is 150062200.

ZT-SCR is a screw part made of aluminum. It is a single axis multi-diameter part with several external and internal features. The external features have five different diameters: one with a diamond knurl, one with a concave surface, one with a tapered surface, and one with threads. The internal features have two internal diameters stepped to one end. The Opitz code found for the screw part is 151001720.

ZT-FLA is a flange part made of nylon. It is a single axis multi-diameter part with both internal and external features. The external features have five different diameters (two pairs have the same diameter) including one with six holes on the pitch circle diameter. The internal features have three different internal diameters stepped to both ends and an internal oil port. The Opitz code found for the flange part is 044052900.

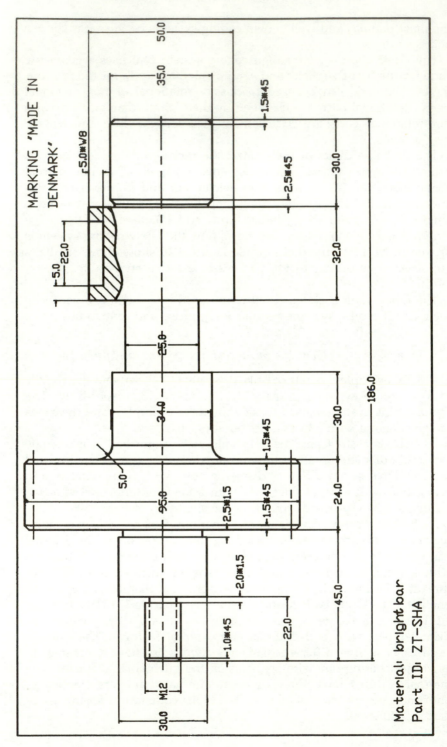

Figure 4.11 An example of a shaft part.

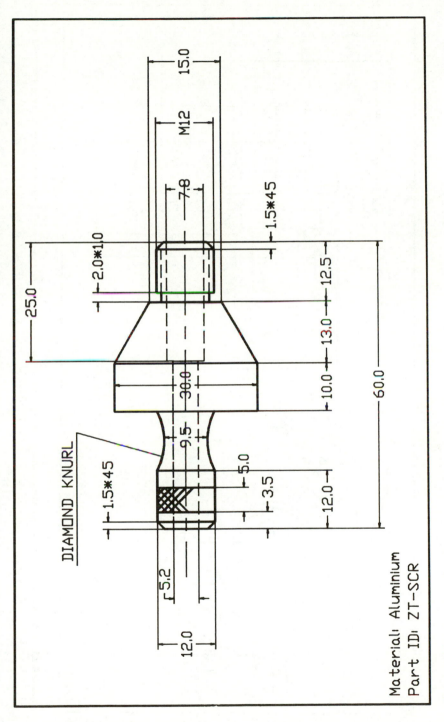

Material: Aluminium
Part ID: ZT-SCR

Figure 4.12 An example of a screw part.

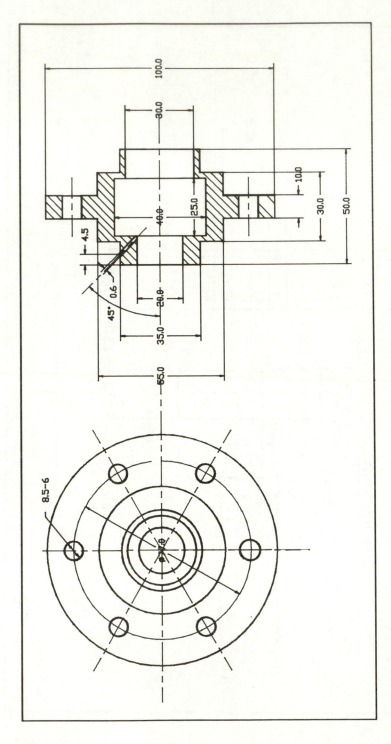

Figure 4.13 An example of a flange part.

REFERENCES

Allen, D.K. and Smith, P.R. (1980) *Computer-Aided Process Planning*. Brigham Young University, Provo, Utah, October 15.

Alting, L. and Lenau, T. (1986) Integration of engineering functions in the CIM philosophy. *Proceedings of ISATA 86*, Flims, Switzerland, October 6–10.

Alting, L. and Zhang, H.C. (1989) Computer aided process planning: the state-of-the-art survey. *International Journal of Production Research*, **27**(4), pp. 553–85.

Chang, T.-C., Wysk, R.A. and Wang, H.-P. (1991) *Computer-Aided Manufacturing*, Prentice Hall, p. 674.

Halevi, G. (1980) *The Role of Computers in Manufacturing Processes*, John Wiley & Sons, New York, p. 502.

Ham, I. (1982) Group Technology. Chapters 7, 8 in *Handbook of International Engineering*, John Wiley, New York. 7.8.1–7.8.19.

Ham, I., Goncalves, E.V. and Han, C.P. (1988) An integrated approach to group technology part family data base design based on artificial intelligence techniques. *Annals of the CIRP*, **37**(1).

Hyer, N.L. and Wemmerlov, U. (1984) Group technology and productivity. *Harvard Business Review*, **62**(4), July/August, pp. 140–8.

Kusiak, A. (1987) The generalized group technology concept. *International Journal of Production Research*, **25**(4), pp. 561–8.

Kusiak, A. (1988) EXGT-S: A knowledge based system for group technology. *International Journal of Production Research*, **26**(5), pp. 887–904.

Kusiak, A. and Chow, W.S. (1987) Efficient solving of the group technology problem. *Journal of Manufacturing Systems*, **6**(2), pp. 117–24.

Milacic, V.R., Uroseviv, M., Veljovic, A., Milev, A. and Race, I. (1987) SAPT – Expert System Based on Hybrid Concept of Group Technology. 19th CIRP International Seminar on Manufacturing Systems, Pennsylvania State University, USA, June 1–2.

OIR Group Technology (1986) OIR Inc. brochures and information material, Lambertus Hortensiuslaan 74 1412, GX Naarden, The Netherlands.

OIR Product News (1987) Advance copy published by Organization for Industrial Research, 100 Crosby Drive, Bedford Ma, USA, January.

Opitz, H. (1970) *A Classification System to Describe Workpieces*, Part I, Pergamon Press, UK.

Teicholz, E. and Orr, J.N. (1987) *Computer-Integrated Manufacturing Handbook*, McGraw-Hill Book Company, New York, p. 444.

Tuffentsammer, K. and Arndt, G. (1983) The influence of NC flexibility on the concept of group technology. *Annals of the CIRP*, **32**(1).

Zhang, H.C. (1987) *Overview of CAPP and Development of XPLAN*. Institute of Manufacturing Engineering, The Technical University of Denmark, Publication No. 87-26.

CHAPTER 5

Automated tolerance analysis for process planning

5.1 INTRODUCTION

5.1.1 Tolerance

Manufacturing parts to exact shapes is known from experience to be impossible. Here the shapes may mean dimensional, geometrical, orientational, or any other characteristics that can be used to describe a feature, or the relationship between features. On one hand, the manufactured parts will be put into diverging environments and their shapes will depart still more from the exact ones, since diverging environments influence the deformation, the wear pattern etc., and consequently also the shapes. On the other hand, the fact that we do not know the exact relationship between the shape of a part and its ability to possess the best possible function makes it neither possible nor necessary to manufacture parts to exact shapes. For example, the torque to be transmitted by a cylindrical interference fit is dependent on the friction between the hole and the shaft. The friction is a function of the interference and other parameters. The value of the interference must be evaluated, based on a specific torque, in designing a cylindrical interference fit. Knowing how complex the contact between metallic surfaces really is, and how difficult it is therefore to predict the value of the friction, one can realize that it is impossible to aim at extreme dimensional precision in this case. Therefore, the variability of the features that can be accepted is the responsibility of the designer. The designer specifies the variation limits of a feature or features, a region within which the features ought to be located so that they retain the functional behavior of the parts. It is important to notice that industry is cognizant of the above understanding and does not specify exact values for features but continuous regions in which no value has any preference. The term 'tolerance zone' is used to refer to this region more precisely, the extreme dimensions of the zones being denoted by the dimension limits. The maximum limit is the greatest acceptable value of the dimension and the minimum limit is the smallest acceptable value of the dimension. Another way of defining the zone is to use the basic (nominal) size and the deviations. The basic size of a dimension is the value from which the calculations of the dimension limits are made. The upper deviation is defined as the difference between the maximum limit and the basic size of the dimension; the lower deviation is correspondingly defined as the difference between the minimum limit and the basic size. The deviation may be positive, zero, or negative as the corresponding dimension limits are greater than, equal to, or less than the basic size. The tolerance zone of a dimension is determined by its size and by its position in relationship to the basic size, and it is defined as the difference between the maximum and minimum dimension limits. Thus, the tolerance is the difference between the upper and lower deviation and is always an absolute value.

The relationship of a tolerance zone to the corresponding feature is given by the location of the datum element from which the measurement is taken. As we have discussed in chapter 2, a datum element can be a surface on a part, but it can also be a center line or a reference plane. The determination of designed datum elements is made by the designer, based on an analysis of the functional requirements of the parts. In the most common cases the design datum elements are not given explicitly. That, for instance, is true for lengths and diameters. In the case of lengths, it is implied that one of the touching end planes is used or that they are mutually datumed. From a manufacturing viewpoint, the end plane of the dimension that is machined first is the datum element of the other one if it is used to locate or measure the position of the other one. In the case of diameters, the center line is implied as the datum element. In more precise and complex cases, the datum elements are given explicitly.

The ANSI standards refer to portions of planar faces as features; if the feature is used as a reference of measurement it is called a datum feature. A cylindrical or spherical surface or a set of two plane parallel surfaces is called a feature of size, each of which is associated with a dimension.

Tolerances can be divided into the following categories:

Size tolerance

A size tolerance is assigned to a physical dimension (excluding position parameters) to state how far individual features may vary from the desired size. Size tolerances are specified with either unilateral, bilateral, or limit tolerancing methods. There are no explicitly specified datums.

Form tolerance

A form tolerance is assigned to a form element (a line, a plane, a circle, or a cylinder) to state how far an actual surface or feature is permitted to deviate from the ideal shape. It is defined as a zone within which the toleranced surface or edge must fit. The zone is bounded by two perfect, offset surfaces or edges; one needs only to specify the offset value and no datum is needed, i.e. the tolerance zone floats in space. The form tolerances include the straightness, flatness, circularity, and cylindricity.

Orientation tolerance

An orientation tolerance also specifies a zone within which a surface or an edge must fit, but a datum is used. That means that the orientation tolerances are the manufacturing variations in the attitude of one feature relative to another. The datum fixes the orientation of the zone but not its position. Sometimes orientation tolerances are applied to a feature of size, like a hole; the implication is to bound the symmetry element (called resolved entity), such as the axis or midplane of the feature, within the tolerance zone. The orientation tolerances include parallelism, perpendicularity, and angularity.

Position tolerance

A position tolerance (formerly called true position tolerances) defines a zone within which the axis or center plane of a feature is permitted to vary from the true (theoretically exact) position. The position tolerance is applied to resolved entities via features of size and up to three datums may be needed. The precedence of the datum must also be specified. Position tolerances include location and coaxiality.

Runout tolerance

A runout tolerance is the composite deviation from the desired form of a revolving part surface, detected during full rotation of the part on a datum axis, either circular of total. The runout tolerance states how far an actual surface or feature is permitted to deviate from the desired form or position specified in order to control the relationship between a datum axis and one or more features. Circular runout specifies permissible variation in length as a part is rotated about a datum axis, i.e. it is a combination of circularity and coaxiality. Total runout controls the entire surface.

Profile tolerance

The tolerance controls the outline or shape of a part as a total surface or planes of a part. The profile of a line tolerance is the condition permitting a uniform amount of profile variation, either unilaterally or bilaterally, along a line element of a feature. The profile of a surface tolerance is the condition permitting a uniform amount of profile variation, either unilaterally or bilaterally, over the entire surface.

Some tolerances require the specification of one or more datums. A datum is an imaginary point, line, or plane used as a reference for measurements. When a part feature is used as a datum it means that the gauge or machine table from which measurement is taking place will be placed such that three points are in contact with the datum feature. When additional datums are needed, the secondary datum must be orthogonal to the primary, and the tertiary orthogonal to the others.

5.1.2 Automated tolerancing in computer aided process planning

A fundamental aspect of design and manufacturing is the transformation of product functional requirements into tolerances of individual components, followed by the production of parts satisfying specified shapes and tolerances. This means that tolerancing is used in two stages in the development of products: design and production. In the design stage, tolerancing assigns tolerances to the shapes of individual parts to ensure that the part can function well in the product and can be produced economically. Usually the designer minimizes the cost of parts by analyzing the cost of manufacturing and assembling. Sometimes he may minimize the cost by accepting some rate of scrap. At present, the cost analysis and estimation are based on the designed dimensions and tolerances (Dong and Soom, 1990; Wu *et al.*, 1988). In the manufacturing stage, the parts cannot all be made in one setup. Therefore, the transformation of datum elements is needed and a dimension and tolerance chain is formed. In this chain, the dimensions may be obtained directly (component dimensions) or indirectly (the sum dimension). Tolerancing is used here to decide the tolerance for each operational dimension to ensure either that the part is made within the designed dimensions and tolerances or that the stock removal for each cut is within certain limits. An operational dimension is the dimension from the datum element, which has a fixed relationship with the machine tool coordinate system and the element to be obtained in the setup. It is clear that the operational tolerances may be equal to or higher (tighter) than the designed tolerances, and only the operational dimensions and tolerances have the direct relationship with the machining process capability, fixturing, tooling, setup, machining time, and scheduling etc. which form the basis for the cost analysis and calculation. Moreover, for those tolerances that have specified datum elements, arrangements of features to be machined to proper setups can heavily influence the accuracy, and consequently the cost, of manufacturing the products. It has been noted that none of the existing CAPP systems can deal with the tolerances with specified datums. This is to say that in order to make parts at the lowest cost, the automated tolerance analysis in the computer aided process planning will play a decisively important role in the future computer integrated manufacturing system (CIMS).

5.2   ASSEMBLY TOLERANCE

The aspect of tolerance technology most commonly discussed in detail is the transformation of functional tolerances given to subassemblies into tolerances of the individual parts concerned, and vice versa. In process planning, the technology is also used in the transformation of the specified tolerance of an indirectly obtained dimension into the tolerances of the relevant individual operational dimensions, and vice versa. Tolerance technology, in this sense of the phrase, can be divided into two groups (Bjorke, 1989; Lee and Woo, 1990):

- Tolerance control
- Tolerance distribution

5.2.1   Tolerance control

The objective of tolerance control is to determine the tolerance of the sum dimension based on given data concerning the individual dimensions in the assembly or in the operational dimension and tolerance chair. Tolerance control is performed in industry in order to verify the functional requirements of a product in cases where the parts exist. It may also be used to check the tolerances found by an approximate tolerance distribution procedure in design and process planning.

The procedure in tolerance control is given below:

1. Determine the component dimensions that influence the sum dimension.

$$D_1, D_2, \ldots, D_n$$

   where D_i is the midpoint of the ith dimension.

2. Determine the fundamental equation of the sum dimension.

$$D_{\text{sum}} = f(D_1, D_2, \ldots, D_n)$$

   where D_{sum} is the midpoint of the sum dimension.

3. Identify or determine the tolerance of the individual dimensions.

$$T_i$$

4. Compute the tolerance of the sum dimension.

$$T_{sum} = sum\{_i\}$$

This tolerance is symmetric to the sum dimension.

5.2.2 Tolerance distribution

The objective of tolerance distribution is to determine, using the economic criteria, the tolerance of the individual dimensions in a chain, based on a specified tolerance of the sum dimension. The objective, in this case, is the opposite of the objective in tolerance control, but the mathematical foundation is the same.

As far as tolerances are concerned, dimensions in a chain may be of two types:

• Dimensions with predetermined tolerances
• Dimensions with determinable tolerances

Tolerances in a chain may be predetermined because the parts may be standard or the dimensions of the parts may be involved in other chains. Examples of standard parts are: ball bearing O-rings, Woodruff keys, V-belts, taper pins, and machine screws. These parts are selected by the product designer and they are included in most mechanical products.

The designer often finds it suitable to his purpose to make rough estimates of tolerances at an early stage in the design procedure. Methods that are useful for this purpose are given below.

Equal tolerances

All of the dimensions having determinable tolerances are given equal tolerances.

$$T = (T_{sum} - sum\{T_{pd}\})/N_{det}$$

where
  T_{sum}: tolerance of the sum dimension
  T_{pd}: predetermined tolerance in the chain
  N_{det}: number of determinable tolerances

Tolerances proportional to the dimensions

Each of the dimensions having determinable tolerances is given a tolerance proportional to the corresponding dimension.

$$T_i = (T_{sum} - sum\{T_{pd}\}) \times D_i/sum_j\{D_j\}$$

where

T_{sum}: tolerance of sum dimension
T_{pd}: predetermined tolerance in the chain
D_i, D_j: dimension having determinable tolerance

Tolerances proportional to the process capability

Each of the dimensions having determinable tolerances is given a tolerance proportional to the process capability for the corresponding dimension.

$$T_i = (T_{sum} - sum\{T_{pd}\}) \times S_i/sum\{S_j\}$$

where

T_{sum}: tolerance of the sum dimension
T_{pd}: predetermined tolerance in the chain
S_i, S_j: process capability for dimension having determinable tolerance

It should be noted that some of the dimensions on a part can be obtained directly from a machining process; their tolerances can be assigned according to the process capability. However, some of the dimensions on the part may be obtained indirectly from a sequence of operations. These dimensions are the sum (resultant) dimensions of the tolerance chains. Their tolerances are the sum of the tolerances of the component dimensions of the chain. When such a dimension is concerned, the tolerance assigned to the dimension should be proportional to the sum of the process capabilities for the component dimensions. In other words, a tolerance should be assigned to each operational dimension proportional to its process capability.

5.2.3   Independence

In the manufacturing of a number of parts, corresponding dimensions will be more or less randomly distributed within the specified tolerance zones. A component of the variability is chance variation; another component is systematic error (for instance, tool wear). Together, these components form the distribution in the population, and it is this distribution the designer has to consider. He cannot turn to account the knowledge about systematic errors in the processes because design and manufacturing are performed in different time intervals.

During assembly, parts are selected randomly from the individual populations and put together. The resulting assembly therefore gives a sum dimension which varies, depending upon the parts selected from the populations. The machining of parts is performed by different machine tools, or by the same machine tool at different points of time. Consequently, the dimensions on the parts are independent. It is important to note that the individual dimensions are stochastic and independent variables, with distributions typical for machining processes used in making the parts, and that the sum dimensions are built up from the individual dimensions, following a functional relationship given by the peculiarity of the assembly. Statistical considerations are therefore valid, even if the number of equal parts is small, because it is the behavior of the machining processes used in making the parts that determines the distributions. The behavior of the machining processes may, on the other hand, be charted and it is this *a priori* knowledge about the statistical behavior of parts that can be and ought to be considered during tolerance calculation.

All of the tolerance analysis in the process of assembly is based on the assumption that the dimensions of the parts are randomly distributed within the specified tolerance zones because the machining of parts is performed in different machine tools, or in the same machine tool at different points in time, and that the parts are selected randomly from the individual populations and put together.

Sometimes assembling with limited interchangeability may be adopted. In direct selection, the assembly fitter selects a pair of mating parts from the population delivered for assembly. During the assembly process, the fitter has to determine if the mating parts give the required function. If the design tolerances are narrower than the capability of the processes to be used to machine the parts, then selective assembly has to be used. Selective assembly is a procedure in which parts are classified into several groups according to size. The parts which are intended to be mated with these are also classified according to size in the same number of groups. Corresponding groups are then expected to assemble and function properly. A verification of the functional requirements is not necessary in this case, since the parts are measured and grouped in a way that gives the desired fit.

The concept of the independence of the part dimensions is very important in tolerance analysis. In the process of assembly with limited interchangeability, the population for each part has been divided and changed. For example, consider an assembly containing a shaft and a bore. The dimensions of the shafts and the bores are 60 ± 0.1 mm and the desired clearance is $0.00-0.02$ mm. If a shaft and a bore are drawn randomly from the two populations, the possible clearance is ± 0.2 mm (worst case). If the distributions of the two populations are similar, most of the parts may be assembled within the specified clearance with limited interchangeability. In this case, the shafts and the bores are divided into subgroups with a tolerance of 0.01 mm. A subgroup of shafts can only fit a subgroup of bores which have the desired clearance. The dimension and tolerance chain theory, which we will discuss in detail later in this book, is only valid for the subgroup of the populations but not for the original populations. This is because, in selection, the dimensions in a fit become interrelated and are not independent any more.

5.3 FUNDAMENTALS OF DIMENSION AND TOLERANCE CHAIN THEORY

The fundamental task of a designer is to transform the functional requirements of products into the dimensions and tolerances of individual components. The fundamental task of a manufacturer is to carry out the production of the components to satisfy the specified dimensions and tolerances. Many, if not all, of the existing tools and techniques developed over the years for tolerance analysis can be embodied in today's computer technology. However, computer software modules in process planning have yet to provide adequate treatment of tolerancing problems, and very few, if any, of them include a systematic check for the feasibility of the process sequence from a tolerance viewpoint. So far there is not a CAPP system which can really generate optimal feasible process sequences. The tolerance information has not been used to produce the setups and sequences but to check a predetermined sequence on a trial-and-error basis. Sometimes the operational tolerances are compressed to less than the process capability to make the components within the designed tolerances. However, an experienced human process planner can make a proper sequence of the operations without compressing the operational tolerances to less than process capacity and still make the components precisely and economically. This means that proper arrangements of features for different setups, according to specified tolerances, are very important for economic manufacturing and should be an integral part of a CAPP system.

5.3.1 Tolerance chain

The theory of the dimension and tolerance (D&T) chain was originally developed for parts assembly because the dimensions of the parts in an assembly are independent. In a simple D&T chain, there is one, and only one, resultant dimension (sum dimension). The relevant dimensions A can be divided into two groups: A_i, whose increases will lead the resultant dimension to increase; and A_j, whose increases will result in the decrease of the resultant dimension. If the resultant dimension is denoted by x,

$$x = \text{sum}\{A_i\} - \text{sum}\{A_j\}$$

Then, in the worst case:

$$x_{max} = \text{sum}\{A_{i\,max}\} - \text{sum}\{A_{j\,min}\}$$

$$x_{min} = \text{sum}\{A_{i\,min}\} - \text{sum}\{A_{j\,max}\}$$

The maximum possible variation of the resultant is:

$$
\begin{aligned}
x_{max} - x_{min} &= \text{sum}\{A_{i\,max}\} - \text{sum}\{A_{i\,min}\} - \text{sum}\{A_{i\,min}\} \\
&\quad + \text{sum}\{A_{j\,max}\} \\
&= \text{sum}\{A_{i\,max} - A_{i\,min}\} + \text{sum}\{A_{j\,max} - A_{j\,min}\} \\
&= \text{sum}\{A_{max} - A_{min}\}
\end{aligned}
$$

That is, in the worst case, the resultant is the sum of the variations of all the relevant independent dimensions.

The theory can be extended to the dimensions and tolerances of a part if each of the dimensions is machined and measured manually or in different setups so that all the dimensions are independent, because each dimension has its own reference element.

5.3.2 Tolerance chart

Used to control tolerance stackups when processing precision machined parts, tolerance chart, a structured methodology, tries to guarantee that all flaws will be purged from the initial process planning decisions responsible for the tolerance stackups. Once this has been done correctly, the resulting in-process cut tolerances can be optimized. The final process will be cost-effective and free of all high-cost tolerance-based problems that are responsible for the generation of scrap or rework on parts, tooling, gauging, and the concomitant elements of frantic debugging and ruptured schedules.

A tolerance chart is based on the theory of the dimension and tolerance (D&T) chain and is a graphical representation of a process plan and a manual procedure for controlling tolerance stackup when the machining of a component involves interdependent tolerance chains (Irani *et al.*, 1989). Such a heuristic, experience-based method of allocating tolerances to individual cuts of a process plan can be embodied in a computer based module.

A tolerance chart recognizes the stochastic nature of dimensions produced on machining cuts, their dependence on machine tool capacity, and the potential effect of combinations of such cuts on resulting dimensions and tolerances. A chart is therefore a graphical representation of the process sequence which allows a process planner to visualize the influence of the proposed sequence on resulting dimensions. It also reduces the problem to a form suitable for tolerance optimization.

At the production stage, planners have the task of transforming raw materials into components that satisfy the specified blueprint form and dimensions. Among the activities typically undertaken is the development of a tentative process plan that will allow for the transformation of raw material shapes into the final component shape (how to develop the process plan using specified tolerances will be discussed later). The process plan is an outline of operations to be performed, datums for setups and their sequence, the machine tool types and tooling involved, and a description of each operation procedure. The feasibility of the tentative process sequence is then checked via a tolerance chart. A tolerance chart provides a graphical representation of each operation obtained in the process plan. Its purpose is to show how individual cuts combine to produce each blueprint dimension.

In the preparation of a tolerance chart, process planners are usually guided by some basic rules:

- The maximum possible tolerance should be assigned to each process cut without violating the specified blueprint tolerances
- Tolerance values assigned to every cut should be consistent with process capability ranges available on the machine
- Minimum and maximum stock removals for each cut should be feasible with respect to the machine tool/tool/work material system under consideration

A number of inputs are required of the process planner for the tolerance chart development. These include the sequence of operations (which will be discussed in detail later), machine tools, the datum element for each setup, surface(s) generated for each operation, and an identification of the operations (or cuts) contributing to each blueprint dimension.

In a tolerance chart, it is assumed that all machining cuts are independent and that a machining cut is represented by a line with an arrow at one end and a dot at the other. A balance dimension is the algebraic combination of two or more cuts and has a dot at both ends. A dot indicates a locating or datum surface, and an arrow represents a newly generated surface. A blueprint dimension is represented by a heavy black line drawn between two machined surfaces. Every blueprint dimension has a corresponding cut if the designed datum coincides with the positioning datum, or with a corresponding resultant dimension on the chart if the two datums do not coincide. Note that all tolerances must be expressed in the equal bilateral system in the tolerance chart. A detailed description of the mechanics of manual tolerance charting can be obtained from Wade (1990).

5.3.3 Tracing (searching) method

Tracing of dimensional chains is carried out backwards from the last operation. Tolerances of final operational dimensions, which are exactly

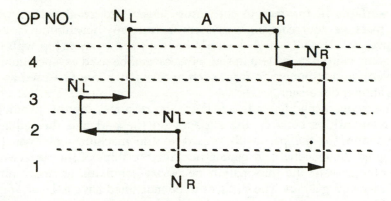

Figure 5.1 Tolerance chain searching.

the same as the dimensions on the part drawing and have no effect on the accuracy of other dimensions in dimensional chains, are taken directly from the part drawing. First, find the two operations machining the two ends of the resultant dimension. All the searches are from arrows machining the ends (dots) of the present dimensions to the ends (dots) of the dimensions with the arrows. The tracing ends when the searches from the two sides of the resultant dimension meet.

For example, in Figure 5.1, dimension A is assumed to be a designed dimension which is either obtained indirectly in a process through some operational dimensions or as an allowance for a certain operation. N_L and N_R are the sequence numbers of the left and right ends of the dimension. Starting from N_L and N_R, searching of the chain is carried out backwards from the last operation. When a dimension in the operation of machining the N_L (or N_R) end is found, the sequence number of the dimension is recorded and N_L (or N_R) is changed into the sequence number of the operational reference of the dimension. Continue this procedure until $N_L = N_R$; the chain searching is then completed. It should be mentioned again here that all the dimensions in a chain should be independent of each other.

In order to understand the following sections, readers need to review the fundamentals of datum theory which we have discussed in section 2.5.

5.4 OPERATIONAL DIMENSION AND TOLERANCE CHAIN

5.4.1 Operational dimensions

An operational dimension is the distance which a surface is machined from a positioning surface in the setup. Because only the positioning surfaces of the part in the setup have fixed relationships with the machine tool coordinate system before any machining is performed, an element which does not have a fixed relationship with the machine coordinate system in the setup cannot be used as a positioning element, and a dimension from such an element cannot be used as an operational dimension.

The sequencing depends on both the accuracy of the geometrical relationships between the surfaces on the part and the required orientations of the part with regard to the machine tool axis. The designed dimensional and geometrical relationships of the surfaces on a part represent the information necessary for fixturing, machining, and measuring a part. The surfaces to be machined have to be arranged in different groups. Each group is machined in a setup and needs a fixture. The number of fixtures must be minimized because of the cost and time involved in the realization of each fixture.

5.4.2   Datum selection

In Chapter 2 we discussed some criteria used for process planning, including datum coincidence and unification. Here we are going to discuss some details of datum selection in computer aided process planning systems.

The selection of the positioning surfaces for a given setup is based on:

- a comparison of the accuracy of the relationships between the already existing surfaces and the surfaces which still have to be machined; this results in the selection of the positioning surfaces.
- the distance between the surfaces which have to be machined and the positioning surfaces.
- the distance between the potential positioning surfaces.
- the orientation and position of a potential positioning surface relative to the remainder of the workpiece.
- the characteristics of the surfaces: size, type, roughness, shape tolerance, orientation, and position with regard to the datum.

The first four items are of primary importance with respect to rotation errors in the position of the part. The last item is important with respect to the selection of the positioning tools.

The machining of a surface results in the specified position of the machined surface with respect to the machine tool coordinate system. This specified position is obtained through the positioning surfaces of the part and the fixture and will be lost if the part is dismounted from the machine tool and remounted in a different fixture. If the positioning surface coincides with the design datum, no tolerance stackups can be introduced because no datum transformation is made.

5.4.3 Setup selection

The alignment error of the fixture on the machine tool or the positioning error of the part to the fixture can be equal to or larger than the accuracy requirements of tight tolerance relationships. As a result, the positional accuracy of a surface which has already been machined in a previous setup can be insufficient to realize the required accuracy in the relations between that surface and the surfaces to be obtained in the present setup. Therefore, the positioning accuracy of fixtures should be analyzed and classified, and closely related surfaces must be machined in one setup if possible. The selection of setups has to be preceded by tolerance analysis, an evaluation of the tolerances concerning the dimensional and geometrical relationships between the different surfaces. After that, the most accurate relationships are selected and the corresponding surfaces become the primary candidates to be arranged in one setup. However, a setup can contain only a limited number of different surface orientations (the maximum number depends on the machine tool configuration). Therefore, only those surfaces are selected whose orientations fit in one setup. In this way, the selected setups result in the minimum requirement for the different fixtures.

5.4.4 Operational dimensions on NC machine tools

Once the workpiece is positioned on the machine tool, it must be correctly aligned with respect to the machine axes, and the datum point of the part program has to be adjusted to correspond with the new position of the component. The most common and easiest way is to arrange the measuring datum (datum of the operational dimension to be machined) to coincide with the locating datum on the NC machine. If the datum of the operational dimension does not coincide with the locating datum, using touch-trigger probes accelerates this otherwise time-consuming activity that is a crucial function on unattended automated machining operations (Smith, 1989). Datum points or surfaces can be found by the probes, and the axes can be automatically zeroed at these locations by the controller. The ability of touch-trigger probes to interrogate the component is almost invariably one of the main reasons for their incorporation on CNC turning and machining centers. However, for the NC machines, which do not have this function, the concept of operational dimensions and tolerances is not the same as that for a manually operated machine tool. In the latter case, the operation is based on the trial-and-error method; therefore, the locating datum may or may not coincide with the measuring datum. On the NC machine, the only dimension that can be achieved directly is the dimension from the datum element on the machine to the surface machined in the setup, unless the measuring datum coincides with the locating datum, or the measuring datum is machined in the same setup. It seems meaningless to give operational dimensions and tolerances without assigning setups and relating the measuring datum to the locating datum. This discussion means that when an operational dimension is given by a CAPP system, it must be related to the machine tool coordinate system, datum elements, setups, and tools as well.

As an example, in Figure 5.2 the operation sequences are typically generated by feature-based CAPP systems (Fainguelernt *et al.*, 1986; Karolin, 1984; Li and Zhang, 1989; Weill, 1988). For Figure 5.2(a), neither are setups clearly assigned nor are operational dimensions given with respect to the coordinates on the NC machines. One cannot expect to perform all the operations on the expensive NC turning centers with touch-trigger probes. The natural locating datum for dimension 4 would be the opposite side of the part. For Figure 5.2(b), an economic tolerance is given to each operation, and the tolerance of the resultant dimension is checked against the tolerance chart method. If the specified tolerance is tight, operation tolerances are compressed to place the resultant tolerance within the specified one (Karolin, 1984; Li and Zhang, 1989). All the previous CAPP systems that deal with tolerance regard the dimensions as independent dimensions. Because of the additive character of the D&T chain, the tolerance of A will be the sum of the tolerances of dimensions 2, 3 and 4. If the tolerance of A is tight, the tolerances of dimensions 2, 3 and 4 will be compressed. Tighter tolerances for the dimensions mean more frequent and accurate calibration or replacement of the tool, lower productivity, and higher accuracy of fixtures and machine tools that lead to longer processing times and higher costs. On an NC machine, if setups and tools are considered, and the tool wear for machining a part is negligible, we can arrange dimensions 2 and 3 in one setup with the same tool. No matter how large the setup errors, tool calibration errors, and the accumulated tool wear, they will not appear in x for any individual part (Figure 5.3). For the part in Figure 5.2, it is not necessary to compress the operation tolerances of dimensions 2 and 3 and use more accurate machines, but rather to arrange them in one setup and use the same tool while only controlling dimension 4. If the four dimensions are performed separately because of machine selection, surface finish, or for some other reasons, and the tolerance of A is still tight, it is also not necessary to compress the tolerances of the operational dimensions as a current feature-based CAPP system does, but rather to add the surface common to dimensions A and B as a new datum element (Figure 5.2(c)). Where the tolerances of B and C are loose, the advantage of the arrangement is obvious. Only one instead of three operational dimensions has a tight tolerance. The information on the

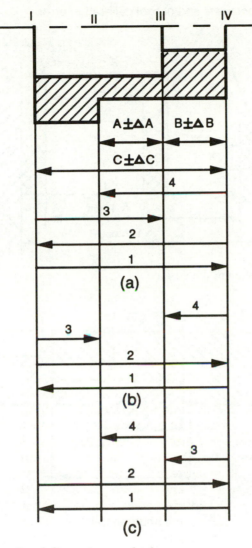

Figure 5.2 Operational dimensions and tolerances.

necessary fixture can also be given at the same time (Figure 5.4). Because some operations cannot be performed separately in two setups, machine selection and scheduling cannot be made without considering the tolerance-based analysis. These examples also show that the method widely used in the assembling dimension and tolerance chain (ADTC) cannot be used directly in the operational dimension and tolerance chain (ODTC).

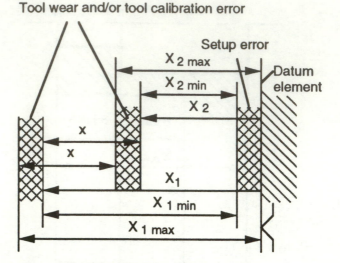

Figure 5.3   Dependent dimensions.

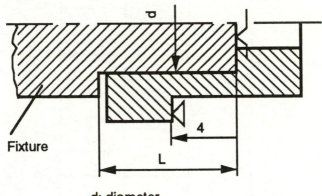

d: diameter
L: minimum length

Figure 5.4   Information on fixtures.

5.5   ALGORITHMS OF AUTOMATED OPERATIONAL SEQUENCING AND TOLERANCING

Many publications have discussed different ways of dealing with tolerancing a rotational part in CAPP. These discussions are based on checking and optimizing the predetermined sequence of a process plan. Here in this section we are not going to review the previous literature; instead, we will focus on some basic concepts and rules for automated operational sequencing and tolerancing.

5.5.1   Orientation

The geometrical requirements of a part are expressed as geometrical relations between the different features. The meaning of a feature requires a particular orientation of the feature with regard to the machine tool axis: the feature orientation. The orientation of a feature is the direction in which the tool can access it. Each feature has at least one default feature orientation. Some features can have more than one feature orientation (e.g. a through hole).

For a rotational part with one axis, all the diameters use the center line as the reference element. Since no datum transformation is concerned when the center support is being used, no chain can be formed for the radial dimensions. Such a part has two orientations for the axial dimensions along the axis. Therefore, all the features on the part can be divided into two groups with different orientations. For the features with only one orientation, two features cannot be machined in one setup if they have different orientations. The identification of orientation can be performed by identifying the adjacent diameters. For example, all of the side surfaces with a larger diameter on one side have the same orientation.

5.5.2   Grouping in setups

A dimension and tolerance chain is formed if the datums for the dimensions concerned are different; therefore, the dimensions are independent. All operations performed in the same setup use the same datum. It is important to group the operations (cuts) with the same orientation into groups that can be performed in one setup. Before grouping the operations, the machining method (turning or grinding) has to be determined. Only the operations that can be performed in the same machine can be put into one group.

5.5.3   Selection of a datum

The diameter dimensions on a part with one axis use the center line as the reference element. When the center support is being used, no datum selection is needed for the diameters. All operations performed in the same setup use the same datums. Usually the datum for the axial dimensions on the part has a different orientation so that it can be located easily on the positioning surface of the machine tool or the fixture to obtain an accurate relationship with the machine tool coordinate system. With a touch-trigger probe on an NC machine tool, a surface with the same orientation can also be used as a datum. It should be noted that the touch-trigger probe has some limitations (see discussion in section 5.6 on Fixture Design), and is expensive and, since it is an in-process measurement, time consuming. It is assumed that all the operations are performed on NC machine tools without touch-trigger probes.

Because all the dimensions machined in the same setup do not form a chain and the operational dimension is equal to the designed dimension if the positioning datum coincides with the designed datum, it is reasonable to select as the datum that element of different orientation which has the tightest designed tolerance with one of the surfaces to be obtained in the setup. If the tolerances are not tight compared with the capability of the process and no preference is given to any of the dimensions, select the plane with the largest diameter in the plane–circle fixturing mode or the longest cylindrical surface in the cylinder–point fixturing mode as the main positioning datum (see Fixture Design in section 5.6).

5.5.4 Sequence

Functionally, the dimensions of a part are non-directional. However, from the point of view of manufacturing, the dimensions called operational dimensions are achieved by a sequence of operations; therefore, each dimension is given a direction from a datum element to the surface being machined. The direct predecessor and direct successor in the chain are the dimension machining the datum element and the dimension using the machined surface as its datum element, respectively, or the dimensions using the same datum elements as depicted in Figure 5.5. If the three operations are arranged in a reverse sequence, a longer chain is achieved with more relevant dimensions.

This example gives us some hints. If a chain must be formed, try to machine the datum in the previous setup and try to reduce the number of component dimensions in the chain. For example, in Figure 5.6 the

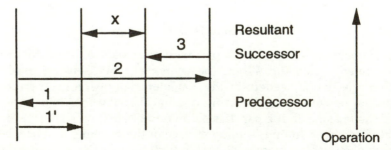

Figure 5.5 Operational dimension and tolerance chain.

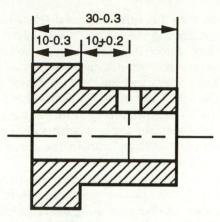

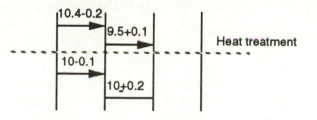

(a)

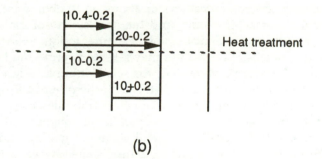

(b)

Figure 5.6   An example of datum and tolerance.

hole on the part has to be machined before the heat treatment and the grinding is performed after the heat treatment. A process planner can make a process plan according to the designed datums, but because of the predetermined operational sequence, the operational tolerances have to be compressed to meet the designed tolerances. If we change the positioning datum for the machining of the hole to the left surface, which can be machined before the machining of the hole, the tolerance chain has only two component dimensions instead of three in the previous plan, and the tolerances for the two dimensions of ±0.1 are twice as much as the 0.1 of the previous plan. This is a typical application of the principle of 'least number' in forming a tolerance chain.

If, in some cases, the datums for two setups are mutually datumed (one is used to machine the other), finish the setup with the tighter tolerance for the resultant dimension last.

By following the rules and procedure described above, a complete operational sequence based on the tolerance-based analysis can be generated with the least number of setups, least number of components in dimensional and tolerance chains, and a maximum possible tolerance for each operational dimension. Therefore, the process plan is economic and optimal. All the algorithms can be implanted in the computer.

5.6 FIXTURE DESIGN

Process planning is one of the bottlenecks in flexible manufacturing of parts in small batches. With reference to present planning methods, the process planner is simultaneously engaged in the selection of machine tools, sequences, setups, tools and fixtures, etc., as we have discussed in the previous chapters. Frequently, a large portion of the time and cost of work preparation is spent on the selection and design of fixtures because they are not integrated with operational sequencing which should be able to generate fixture information according to the dimensional and geometrical specifications on the blueprint, restraints of capability of machine tools, and the accessibility of tools. Here we devote a section to a brief discussion of fixture design. Readers who are interested in details of the subject may refer to Boerma and Kals (1988).

The use of a generative system for selection and design of fixtures will only be efficient if it is integrated with the other process planning functions. The fixturing process function (module) uses the fixture and setup information given by operational tolerance analysis and sequence, selects the positioning and clamping modes, and examines the fixture database to see if such a fixture is already available. If not, it automatically produces all of the parameters for the fixture and installs them into the database.

The primary functions of a fixture are to position, to clamp, and to support the part. Positioning of a part is based on the so-called 3-2-1 method to retain the six degrees of freedom. The first datum plane can restrict three degrees of freedom (two rotations and one translation). The second datum plane can eliminate two degrees of freedom (one rotation and one translation). The third datum will remove the last one. It is assumed that the part has at least a 3-point (not on a line) contact with the first datum plane, and at least a 2-point contact (not on a line normal to the first datum plane) with the second datum plane.

The geometrical requirements of a part are expressed as geometrical relations between the different surfaces. The machining of a surface requires a particular orientation of the feature with regard to the machine tool axis.

5.6.1 Type

The fixtures used most often in turning processes are 3-jaw (automatic centering) chucks, flange-type fixtures, and center-supports.

3-Jaw chuck

Geometrically a chuck can restrict five degrees of freedom, and the clamp will remove the last degree of freedom of the part relative to the chuck. If a cylindrical surface and another element (a point or a plane) are used as positioning elements, the center line of the part can theoretically coincide with that of the spindle. If the contacting length of the cylinder with the jaws is long enough, four degrees of freedom can be removed and only one point is needed to locate the part in the z direction. Otherwise a plane has to be used as a positioning surface which will restrict three degrees of freedom and the cylinder, which now can only be regarded as a circular element, will remove two degrees of freedom.

Flange

A flange-type fixture functions like a chuck, having been adjusted to a certain dimension, except that the clamping mechanism may be different. The positioning mode can also be a cylinder–point type or a plane–circle type.

Center-support

In center-supported mode, the part is supported by two cones at the center holes or chamfers on the two ends of the part. A clamp or equivalent is used to transmit the torque for rotation.

5.6.2 Datum

Surfaces which are dealt with in a given setup can have explicitly defined geometrical relationships with the surfaces which have to be machined in the next setup. Within a setup, existing surfaces carry the most important relationships with those which still have to be machined, and serve as positioning surfaces which have to be located at the prescribed positions of the machine tool coordinate system. The surfaces which are actually used to position the part are the so-called positioning surfaces.

5.6.3   Error analysis

3-Jaw chuck

The three jaws are driven by a plate with an Archimedes curve. It is so designed and manufactured that the three locating elements on the jaws can be adjusted simultaneously and always define a circle with the center fixed at the center of the chuck. The errors vary with the accuracy with which the chucks are made and assembled and how they are, and have been, used. Because clearances should be kept between the jaws and the chuck to ensure free movement of the jaws, the jaws may rotate a little when clamped. The rotations of the jaws may not be equal. Therefore, in a plane–circle positioning mode, the positioning plane formed by the jaws may not be parallel to the generating plane. This error will directly influence the dimension between the two planes. The translation errors resulting from inadequate alignment can be compensated by the machine tool control system, which means that they need to be considered in the planning procedure. Contrary to this, rotation errors cannot be compensated. Therefore, the required relationship can best be realized when the corresponding surfaces are machined in one setup. This is the reason why some surface elements have to be machined in one setup if the dimensional or positional tolerance is tight. Of course, this may not be possible when the surfaces have different orientations.

Flange type

Most of the positioning errors of the parts come from the positioning error of the fixture relative to the machine tool coordinate system and the clearance between the cylindrical surfaces of the part and the fixture, since the diameter of the cylindrical surface of the fixture is fixed. If the fit between them is tight (interference fit), high concentricity may be obtained, but the tolerance of the diameter on the part may be tighter than specified on the blueprint, and it may be difficult to put the part into the fixture and to take it out of it. If the fit is loose (clearance fit), the non-concentricity may be as large as the clearance. Therefore, a transition fit is recommended, and the maximum radial clearance between the part and the fixture has to be checked to make sure that the part is made within the specified concentricity requirement. Because the clamp force can be applied along the z direction, only the positioning error of the fixture relative to the machine tool is responsible for the positioning error of the part in that direction.

Center-support

The positioning repeatability of a center-support is good for most of the relative relationships among the featurces. The features may use the center holes repeatedly in different setups as if they were in one setup. If the part changes its direction in two setups, the axial dimensions may form a chain because the datum for the z direction has been changed.

5.6.4 Clamping

The clamping function is performed by locking the part to the positioning surfaces by means of one or more clamping tools which may or may not be the positioning elements themselves. During the clamping process, the position of the part has to be determined by the positioning elements and must not be influenced by the clamping forces. Positioning surfaces are the first candidates to become primary supporting elements. Subsequently, the clamping surfaces are, as much as possible, selected at locations opposite the supporting surfaces. The common clamping modes for flange type fixtures are axial and radial (expanding).

5.6.5 Parameters

The outputs of operation sequencing for fixturing are the dimensions and tolerances of the positioning surfaces of the required fixture. The function of a fixturing system is to check if a chuck or a flange type fixture is available whose dimensions, capability (accuracy), and accessibility are suitable for that part. For a flange type fixture, the clamping mode has to be checked if the machining can be free from interference from the clamping. If not, the system will automatically generate a flange type fixture by means of which the part can be mounted accurately and reliably and made within the designed specifications. The outputs of the system are the dimensions and tolerances of the positioning surfaces of the fixture, the clamping mode, and the position relative to the datum on the machine to which it is mounted.

5.7 TOLERANCE BASED PROCESS PLANNING

Among the many tasks required of process planning are the specifications of the sequence of operations, machine tools, cutting tools, and
fixtures necessary to transform the raw material form into a shape
satisfying the dimensions and tolerances appearing on the blueprint
and the estimated cost of manufacturing. A tentative process sequence
developed by a process planner, or by a computer software module,
must ultimately ensure that dimensions and tolerances specified on
the design are satisfied without undue cost or difficulty. A human
process planner with little knowledge cannot produce good process
plans; neither can a CAPP system. Without the tolerance analysis, the
sequence of operations and setups cannot be obtained, because the
sequence of the setups will influence the tolerance stackups directly.
Without the analysis, the machine tools cannot be selected and the cost
estimation cannot be made, because it is the operational tolerances
rather than designed tolerances that are closely related to the machine
tool capability. Without the analysis, the fixtures cannot be decided,
since the fixtures are selected or designed according to the setups.
Without the analysis, scheduling cannot be made, since alternatives
cannot be generated. All the functions mentioned above constitute the
main tasks of CAPP systems, and none of them can be done without,
or prior to, the tolerance analysis.

In our view, automated tolerance analysis significantly compacts the
entire procedure of process planning. Without tolerance analysis, process planning cannot be done or the process plans cannot be executed
in the real manufacturing environment. CAPP systems cannot really be
integrated into the real manufacturing environment unless the CAPP
systems can perform tolerance analyses. This discussion draws our
attention towards tolerance based process planning, which may be
considerably different from current existing process planning systems.

REFERENCES

Bjorke, O. (1989) *Computer-Aided Tolerancing*, ASME Press, New York.
Boerma, J.R. and Kals H.J.J. (1988) FIXES, a system for automatic selection of
  set-ups and design of fixtures. *Annals of the CIRP*, **37**(1), p. 443.
Dong, Z. and Soom, A. (1990) Automatic optimal tolerance design for related
  dimension chains. *Manufacturing Review*, **3**(4), December, p. 262.
Fainguelernt, D., Weill, R. and Bourdet, P. (1986) Computer-aided tolerancing
  and dimensioning in process planning. *Annals of the CIRP*, **35**(1), p. 381.
Irani, S.A., Mittal, R.O. and Lehtihet, E.A. (1989) Tolerance chart optimization.
  International Journal of Production Research, **27**(9), p. 1531.
Karolin, A. (1984) Computer aided tolerance analysis. *AUTOFACT* **6**, MS84–
  762.

Lee, W.J. and Woo, T.C. (1990) Tolerances: their analysis and synthesis. *Journal of Engineering for Industry*, **112**, p. 113.

Li, J.K. and Zhang, C. (1989) Operational dimensions and tolerances calculation in CAPP systems for precision manufacturing. *Annals of the CIRP*, **31**(1), p. 483.

Mallur, S., Mei, J. and Zhang, H.C. (1992) An Introduction to an Integrated Process Planning Model, International Manufacturing Conference – 92, Dallas, Texas, March 30–April 2.

Mittal, R.O., Irani, E.A. and Lehtihet, E.A. (1990) Tolerance control in the manufacturing of discrete components. *Journal of Manufacturing Systems*, **9**(3), p. 233.

Shah, J.J. and Miller, D. (1989) A Structure for Integrating Geometric Tolerances with Form Features and Geometric Models. *Proceedings of ASME Conference on Computers in Industry*, p. 395.

Smith, G.T. (1989) *Advanced Manufacturing, the Handbook of Cutting Technology*, IFS Publications, UK.

Wade, O.R. (1990) The Role of Tolerance Stack-up Control in Manual or CAPP systems, Pacific Conference on Manufacturing, Australia, p. 1252.

Weill, R. (1988) Integrating dimensioning and tolerancing in computer aided process planning. *Robotics & Computer-Integrated Manufacturing*, **4**(1/2), p. 41.

Wick, C. and Veilleux, R.F. (1988) *Dimensional Metrology and Geometric Conformance*, SME.

Wu. Z., Elmaraghy, W.H. and Elmaraghy, H.A. (1988) Evaluation of cost-tolerance algorithms for design tolerance analysis and synthesis. *Manufacturing Review*, **1**(3), p. 168.

Dimensioning and Tolerancing, American National Standards Institute (ANSI) Standard ANSI Y14.5M.

CHAPTER 6

Computer aided process planning

6.1 INTRODUCTION

Computer Aided Process Planning (CAPP) can be defined as the functions which use computers to assist the work of process planners. The levels of assistance depend on the different strategies employed to implement the system. Lower level strategies only use computers for storage and retrieval of the data for the process plans which will be constructed manually by process planners, as well as for supplying the data which will be used in the planner's new work. In comparison with lower level strategies, higher level strategies use computers to automatically generate process plans for some workpieces of simple geometrical shapes. Sometimes a process planner is required to input the data needed or to modify plans which do not fit specific production requirements well. The highest level strategy, which is the ultimate goal of CAPP, generates process plans by computer, which may replace process planners, when the knowledge and expertise of process planning and working experience have been incorporated into the computer programs. The database in a CAPP system based on the highest level strategy will be directly integrated with conjunctive systems, e.g. CAD and CAM. CAPP has been recognized as playing a key role in CIM.

More than 20 years have elapsed since the use of computers to assist process planning tasks was first proposed. Tremendous efforts have been made in the development of CAPP systems. A few general survey works have been reported in the last decade (Weill *et al.*, 1982; Chang and Wysk, 1985; Wysk *et al.*, 1985; Eversheim and Schulz, 1985; Ham and Lu, 1988; Alting and Zhang, 1989). In this chapter the historical development of CAPP will be reviewed. The necessity of Automated Process Planning (APP) will then be discussed. Finally, three distinctive approaches to CAPP: the variant approach, the semi-generative approach, and the generative approach will be described.

6.2 HISTORICAL BACKGROUND

In 1965 Neibel first presented the idea of using the speed and consistency of the computer to assist in the determination of process plans. Next, Schenk discussed the 'feasibility of automated process planning' in his Ph.D. dissertation at Purdue University (Schenk, 1966). Despite the early recognition of the possibility of extracting operations and processing sequences from the part geometry described in CAD, computer aided process planning was not broadly addressed until the beginning of the 1970s. This was probably due to the fact that the computer capabilities of both hardware and software were limited and that manufacturing engineering was somewhat isolated from computer aided techniques at that time. 1976 was probably the first 'harvest' year in the CAPP area. The CAPP (an actual system name) System, the first variant system, was developed under the direction and sponsorship of CAM-I (Computer Aided Manufacturing-International) and presented at the 1976 NC Conference (Link, 1976). In the same year another system, MIPLAN, developed by the OIR (Organization of Industrial Research), was presented (Houtzeel, 1976). Since then CAPP has been widely addressed. Subsequently, in 1977, Wysk presented a generative system for detailed process selection called APPAS in his Ph.D. dissertation (Wysk, 1977). In 1978, DCLASS, which stands for Decision Classification Information System, was developed by the Computer Aided Manufacturing Laboratory at Brigham Young University and first applied and installed at Boeing Aircraft for the purpose of generative planning for aircraft parts (Allen and Smith, 1980).

By the late 1970s and the early 1980s, Artificial Intelligence (AI) techniques were introduced and applied in the CAPP area. Many CAPP systems were implemented using AI techniques, which were usually entitled 'Knowledge-based' or 'Expert' systems. Some fundamental AI techniques relevant to CAPP have already been discussed in Chapter 3. In 1981, GARI, the first CAPP system that addressed AI techniques, was developed at the University of Grenoble in France (Descotte and Latombe, 1981).

From 1985, researchers in the development of CAPP have put their efforts on a higher level. While the researchers in industry focused on introducing variant CAPP systems into a real manufacturing environment, the researchers in academia focused on introducing AI concepts into the development of generative CAPP systems. Simultaneously, some other research institutions focused their attentions on integrated CAPP systems in conjunction with manufacturing functions. By the end of the 1980s, many intelligent and integrated process planning systems had appeared. In 1987, the International Institution for Production Engineering (CIRP) hosted an international seminar on the main themes of CAPP in Pennsylvania State University, USA. More than 100 technical papers represented about 40 CAPP systems. Some intelligent and integrated process planning systems were reported at the seminar, such as XMAPP (Inui *et al.*, 1987), KAPPS (Iwata and Fukuda, 1987), SAPT (Milacic *et al.*, 1987), TOJICAPP, BITCAPP and BHCAP (Jiang and Xu, 1987), EXCAD and ICAPP (Wright *et al.*, 1987). Continuing, in 1989, CIRP hosted an international workshop on the theme of CAPP, in Hanover, Germany. Some new integrated CAPP systems were reported, such as PART (Houten *et al.*, 1989), DPP (Iwata and Fukuda, 1989), and FLEXPLAN (Tönshoff *et al.*, 1989). These systems, focused on from a research point of view, discussed the integration of process planning and production scheduling. Dr. Iwata provided the concept of a dynamic process planning system, while Professor Tönshoff presented a non-linear process planning system. With the same goal of the integration of process planning and scheduling, DPP emphasizes dynamic planning, based on the real-time information from the shop floor, while FLEXPLAN generates alternative process plans which can be selected by scheduling functions. This contribution has pushed the research interests into an integration of process planning with shop floor control, which is one step closer to integrated production. In 1990, CIRP hosted an international seminar again on the theme of CAPP, in the Netherlands. While some new systems were reported, more technical papers focused on detailed technical problems. Spur provided a presentation for advanced software tools for the development of CAPP systems (Spur *et al.*, 1990). Lu emphasized a conceptual model for automated process planning for the machined parts domain (Ferreira *et al.*, 1990). Chang discussed a new approach for geometric reasoning into integrated process planning (Chang, 1990). This conference may be the point from which researchers in the development of CAPP systems move their attentions to concurrent or simultaneous engineering. From the beginning of the 1990s until the present, few new systems have been reported but many new concepts, methodologies, and technical issues have been addressed. To date, about 187 individual CAPP systems based on different implementation approaches have been reported and investigated. Some of them are well known systems and will be discussed in Chapter 9.

For the time being, the research interests for development of CAPP systems are focused on intelligent and integrated process planning systems. For increasing the intelligence of CAPP systems, some new concepts, such as neural networks, fuzzy logic, and machine learning have been explored for the new generation of CAPP systems. For increasing the integrability of CAPP systems, feature based design, the roles of features, integrating process planning with scheduling, and integrating process planning with manufacturing resources planning have been focused on. This phenomenon is entitled concurrent or simultaneous engineering.

6.3 WHY COMPUTER AIDED PROCESS PLANNING?

It is obvious that CAPP development has been addressed by many universities, institutions, research organizations and corporate development departments. A great effort has been made on the subject. However, the question of why CAPP is so important for the current production environment still needs to be answered. In this section the issues will be addressed. In general, there are three main arguments that are involved in the subject.

Since a process plan determines the methods, machines, sequences, fixturing, and tools required in the fabrication and assembly of components, it is easy to see that process planning is one of the basic tasks to be performed in manufacturing systems. The task of carrying out the difficult and detailed process plans has traditionally been done by workers with a vast knowledge and understanding of the manufacturing process. Many of these skilled workers, now considered process planners, are either retired or close to retirement, with no qualified young process planners to take their place. An increasing shortage of process planners has been created. With the high pressure of serious competition in the world market, integrated production has been pursued as a way for companies to survive and succeed. Automated process planning systems have been recognized as playing a key role in CIM. It is for reasons such as these that many companies look for computer aided process planning systems.

Computer aided process planning is the way in which most companies are solving the problem of automating process planning and overcoming the shortage of skilled process planners. As the American Machinist and Automated Manufacturing Society has reported in the paper Process Planning Software Enhances Accuracy and Consistency, a computerized process planning system has essentially four goals: (1) reduce the clerical load of plan preparation on the manufacturing engineers and skilled process planners, who are in short supply; (2) optimize existing plans using the best available information on machines, tools, speeds, etc.; (3) standardize what are known to be the 'best' process plans for families of components within a company, thereby capturing the knowledge of the skilled planners; and (4) standardize production times/costs for particular families of components.

6.3.1   Requirement of consistency

The process planner's task has been described in section 2.5. It is complex work and involves tremendous working experience. Usually a good process plan is carried out by a sophisticated process planner. He or she is expected not only to have a high standard of formal education in engineering, but also to have considerable workshop experience, possibly as a machinist. Each company has its own rules and know-how for process planning and the facilities and equipment are different for each company. In order to carry out a well-suited process plan for a specific company, a certain amount of practical experience in that company is always required by the process planner. Even so, it is quite often that an identical part is planned ambiguously by different process planners. Furthermore, sometimes the same process planner designs various plans over a period of time for identical parts. The inconsistency of manual process planning is both time consuming and tedious. Carrying out an inconsistent plan will cause extra tasks in production planning, management, and resource planning. Sometimes equipment which has already been made ready for use with the consistent plan has to be rearranged again for an inconsistent plan. A great deal of duplicate work cannot be avoided for the inconsistent plan and may even cause some extra purchasing of tools and fixtures which may be substituted for the consistent plan's fixturing.

6.3.2   Process planners shortage

In spite of the drawbacks of the manual process planning approach, most tasks of process planning are still performed by skilled process planners. However, in recent years, declining numbers of skilled process planners have been reported in some industrialized countries, such as the USA (Emerson and Ham, 1982; Steudel, 1984), UK (Davies *et al.*, 1986), etc. Many skilled process planners are reaching retirement age, with fewer qualified people to replace them. Companies are facing two emerging problems:

- To gather and save the expertise (the knowledge and working experiences of process planning) from those process planners, so that the companies can still use the expertise after they leave.
- To compensate for the shortage of skilled process planners by means of computer systems which can take a certain amount of tasks from process planners.

These problems give impetus to the requirement of CAPP in manufacturing companies.

6.3.3 Requirement of integration

With the rapid development of Computer Aided Techniques (CAT), both CAD and CAM have developed advantageously. Unfortunately CAD and CAM are developed separately. There is an obvious gap between CAD and CAM – namely manual process planning. The level of technological development in process planning is clearly lagging, which is a major deterrent to achieving goals in productivity improvement.

The potential benefits of increasing productivity and flexibility cannot be fed back unless the integration of CAD and CAM is achieved, in which tremendous investment has been made. The integration of CAD and CAM cannot be achieved unless computer aided process planning can match the development of CAD and CAM. From this statement one can see that process planning becomes a bottleneck in the manufacturing activity. In order to realize fully integrated production, automated process planning must be reached.

6.4 COMPUTER AIDED PROCESS PLANNING APPROACHES

In general, two approaches to CAPP are traditionally recognized: the variant approach and the generative approach. However, with the rapid development of new techniques, many CAPP systems do not exactly fit this classification and combine both approaches, so that a third category is now recognized, the semi-generative approach. In this section these three approaches will be discussed.

6.4.1 The variant approach

The variant approach to process planning is comparable with the traditional manual approach where a process plan for a new part is created by recalling, identifying, and retrieving an existing plan for a similar part (sometimes called a master part), and making the necessary modifications for the new part. In some variant systems parts are grouped into a number of part families, characterized by similarities in manufacturing methods and thus related to group technology. For each part family, a standard process plan, which includes all possible operations for the family, is stored in the system. Through classification and coding, a code is built up by answering a number of predefined questions. These codes are often used to identify the part family and the associated standard plan. The standard plan is retrieved and edited for the new part. The variant approach is widely used, e.g. CAPP (a real computer aided process planning system) (Link, 1976), MIPLAN (Houtzeel, 1980), etc. In comparison with manually performed process planning, the variant approach is highly advantageous in increasing information management capabilities. Consequently, complicated activities and decisions require less time and labor. Also, procedures can be standardized by incorporating a planner's manufacturing knowledge and structuring it to a company's specific needs. Therefore, variant systems can organize and store completed plans and manufacturing knowledge from which process plans can be quickly evaluated. However, there are difficulties in maintaining consistency in editing practices and in the inability to adequately accommodate various combinations of geometry, size, precision, material, quality, and shop loading. The biggest disadvantage is that the quality of the process plan still depends on the knowledge background of a process planner. The computer is just a tool to assist in manual process planning activities. However, the variant approach is still popular. The main reasons probably are:

- The investment in hardware and software is less. Vendors for variant systems are more available now as compared with generative systems.
- The development time is shorter and manpower consumption is lower. Installation is easier than for generative systems.
- In the current situation, the variant system is somewhat more reliable for use in real production environments, so it is reasonable for current production environments, especially for small and medium sized companies which do not have resources to form a research/ development group, to choose this approach. An example of a variant system is MIPLAN, which was developed in conjunction with the Organization for Industrial Research (OIR) (Houtzeel, 1976). MIPLAN uses the MICLASS coding system for part descriptions. They are data retrieval systems which retrieve process plans based on part code, part number, part matrix, and code range.

6.4.2   The generative approach

The highest level of automation and sophistication in computer aided process planning is the generative approach. As the name implies, the approach takes the design specifications and turns them over completely to the computer and its programming for process plan production. In the generative approach, process plans are generated by means of decision logic, formulae, technology algorithms, and geometry based data, to perform uniquely the many processing decisions for converting a part from raw material to a finished state. The rules of manufacturing and the equipment capabilities are stored in a computer system. When using the system, a specific process plan for a specific part can be generated without the involvement of a process planner. For generative systems, input can come either as a text input, where the user answers a number of questions in an English or English-like dialogue (defined as interactive input), or as graphic input, where the part data is gathered from a CAD module (defined as interface input). The former is more common in existing CAPP systems, while the latter is still a fairly undeveloped area due to its complexity. Nevertheless, interface input is necessary to permit an integrated manufacturing system. It has attracted much effort in an attempt to interface CAPP with CAD. The terms feature recognition, feature extraction, feature refinement, and geometry reasoning have been used to denote this study, which will be discussed in detail later in this book. Tulkoff states that 'Generative process planning systems today are still somewhat elusive, on the whole, and can be considered as being in their early stages of development and use' (Tulkoff, 1987). The generative approach is complex and a generative CAPP system is difficult to develop. At the beginning some argued that this type of system was too complex to ever be computerized. However, the rapid development of AI techniques and the successes of applying AI techniques in other areas have greatly encouraged the utilization of AI techniques in process planning. Process plans produced in this manner are consistent, fully automated, and may be completely integrated with computer integrated manufacturing. Results of such systems are impressive; the US Navy reports that it has reduced lead times from 300 to 30 days for many repair parts using a generative system (Ohr, 1990). True generative systems are still in their infancy, however, and are therefore extremely expensive, time consuming to implement, and applicable to only a small range of the parts that must be machined. This effort has given initial results which indicate that generative systems are desirable and promising.

Several generative process planning systems have already been developed such as APPAS (Wysk, 1977), CMPP (Waldman, 1983), EXCAP (Davies and Darbyshire, 1984), XPLAN (Lenau and Alting, 1986), and so on. The biggest advantage of the generative approach is that the process plan is consistent and fully automated. This kind of system is mostly oriented towards large companies and research organizations, since they can afford the investment of a long term project. For companies which have a number of products in small lot sizes, the generative approach is particularly attractive. As a research field to enable the necessary integration within the CIM concept, the generative approach is important. Five alternative approaches to generative process planning are discussed in detail by Allen (1987):

- Decision tables
- Decision trees/decision tables
- Axiomatic
- Rule-based decision tree
- Constraint-based

A good example of how a generative system can save time and money is shown in an example Chevalier gave when he went to analyze a client that produced 523 different gears, and process planners had developed 477 different process plans for these gears. 'Since each plan averaged about three pages, there were nearly 1500 pages of process plans in the files. A process planner faced with the development of a plan for a new gear was not likely to contend with 1500 pages of process plans to determine the appropriate way of producing the new gear. As a result, new plans were prepared for each new gear. After analysis, we found that more than 400 of the process plans could be eliminated. When we consider that each plan took 2 to 3 hours to prepare, it is obvious that a significant saving would result from simplifying the entire situation, so that new process plan development would be minimized.' (Chevalier, 1983)

An example of a generative system is CAPP, which is an acronym for CAM-I's Automated Process Planning System, and which was developed by McDonell Douglas. CAPP was developed as a research tool to demonstrate the feasibility of computer aided process planning. It is structured for a database, retrieval logic, and interactive editing capability (Alting and Zhang, 1989).

'The major advantages of generative process planning are the rapidity and consistency with which plans may be generated and ease of incorporating into the plans new processes, equipment, methods, and tooling.' (Allen and Smith, 1980) Generative systems are attractive for companies with a number of products in small lot sizes.

6.4.3 The semi-generative approach

The semi-generative approach is an interim approach and it is still in its infancy. Emerson and Ham, when they presented a semi-generative system called ACAPS, stated that 'It must be said at this point that the purely generative CAPP system has yet to be developed. Until such time as a generative system emerges, much effort has gone into semi-generative CAPP systems. These serve to reduce user interaction through such features as standard operation sequences, decision tables and mathematical formulas. These schemes are not completely generative, but they can be extremely useful in terms of time and cost savings in the manufacturing environment' (Emerson and Ham, 1982). In order to highlight the differences between it and the final approach, the semi-generative method entails interaction between a computer and a human, who has some degree of expertise in the area of process planning. The planner's responsibility is the interpretation of design data and/or a mechanical drawing. The computer prompts the operator for measurements, materials and tolerances, and produces a process plan based upon predefined algorithms and formulae. The term 'semi-generative' approach may be defined as a combination of the generative and the variant approaches, where a pre-process plan is developed and modified before the plan is utilized in a real production environment. It means that the decision logic, formulae, and technological algorithms, as well as the geometry based coding scheme for translating physical features (such as features' sizes and tolerances, locations and surface roughness, etc.), are built into the system. At first sight, the system's working steps are the same as for the generative approach, but the final process plan has to be examined and errors corrected if it does not fit into the real production environment. It may be a good idea to break a generative system down into a plan generating stage and a modifying stage to correct the plans which may be in conflict with the specific production environment. Modifying is small compared with the variant approach. From a research point of view the semi-generative system may not be the desired direction, but it increases the system's competitiveness on the market. Industrial application of such systems can (1) speed up automatic production, (2) reduce the process planner's participation, and (3) ensure the quality of the process plan. Since it is a practically oriented system for industry, the semi-generative approach may be a good candidate during the transition period.

There are some other approaches, such as the constructive approach (Lyons, 1986) and the artificial intelligence approach, but these approaches can be included in one of the above three categories.

To summarize, the advantages of CAPP are typical of those accrued when any procedure is automated via computers. A brief list of these advantages are:

- reduced clerical effort
- fewer calculations
- fewer oversights in logic
- immediate access to up-to-date information
- consistent information
- faster response to engineering or production changes
- use of latest revisions
- more detailed and uniform planning
- more efficient use of resources

6.5 NEW GENERATION OF CAPP SYSTEMS

In almost the last three decades, the aspect of CAPP has been dramatically changed. Although the final goal of CAPP research remains in the same direction, its contents and emphases have gone through significant changes during the time period. Many new generation CAPP systems have been developed recently. In comparison with traditional CAPP systems, the new generation of CAPP systems have several advantages. First, artificial intelligence (AI) techniques have significantly impacted the development of CAPP systems. Although we still consider that the new generation of CAPP can be categorized into the three approaches, variant, generative, and semi-generative, AI-based CAPP systems are remarkably different from traditional generative CAPP systems. The implementation tools for the new generation systems have involved many new techniques, such as knowledge based techniques, object-orientated programming techniques, common product model, and virtual single manufacturing database techniques. In terms of the application of AI techniques in the development of CAPP, not only knowledge base and expert systems are used, but also fuzzy logic and neural network techniques have been involved. Some new generation systems have employed the machine learning approach (Lu and Zhang, 1990). The second difference in comparing new generation CAPP systems with traditional CAPP systems is that the integrability has been dramatically improved. As we have discussed in the previous chapters, CAPP plays a key role in integrating design and manufacturing. In terms of the integration of design and manufacturing, the feature techniques have been recognized as essential tools for eventually integrating process planning and design. Many researches have resulted in some applicable approaches such as feature recognition, feature classification, geometrical reasoning, etc. Many feature based process planning systems have been reported recently. The feature technique is so important that we devote a particular chapter (Chapter 8) to its discussion. In terms of integration with other functions, there are also some good results reported for the integration of process planning and production scheduling. So far, researches in this area have utilized several approaches in terms of nonlinear process planning, dynamic process planning, closed loop process planning, just-in-time process planning, and so on. Although the final integration of process planning with production scheduling is still on its way, some initial research has resulted in quite promising progress. We will particularly discuss the integration of process planning and production scheduling in Chapter 10.

Generally speaking, the difference between the new generation of CAPP and traditional CAPP lies in three aspects: (1) integrability, (2) intelligence, and (3) high techniques orientation. Several prototype new generations of CAPP systems which can be pointed out are QTC, FLEXPLAN, PART, etc. We will discuss these systems in Chapter 9.

REFERENCES

Allen, D.K. and Smith, P.R. (1980) *Computer-Aided Process* Planning. Brigham Young University, Provo, Utah. Oct. 15.

Allen, D.K. (1987) An introduction to computer-aided process planning. *CIM Review*, Fall.

Alting, L. and Zhang, H.C. (1989) Computer aided process planning: the state-of-the-art survey. *International Journal of Production Research*, **27**.

Chang, T.-C. and Wysk, R.A. (1985) *An Introduction to Automated Process Planning Systems*. Prentice-Hall Inc., Englewood Cliffs, New Jersey, p. 230.

Chang, T.-C. (1990) Geometric Reasoning – The Key to Integrated Process Planning. Proceedings of the 22nd CIRP International Seminar on Manufacturing Systems, University of Twente, Enschede, Netherlands.

Chevalier, P.W. (1983) Group Technology: The Connecting Link to Integration of CAD and CAM. Autofact Europe Conference.

Davies, B.J. and Darbyshire, I.L. (1984) The use of expert systems in process-planning. *Annals of the CIRP*, **33**(1).

Davies, B.J., Darbyshire, I.L. and Wright, A.J. (1986) The Integration of Process Planning with CAD CAM Including the use of Expert Systems. Proceedings of the International Conference on CAPE, Edinburgh, UK, April.

Descotte, Y. and Latombe, J.-C. (1981) GARI: A Problem Solver that Plans how to Machine Mechanical Parts. Proceedings of the 7th International Joint Conference on Artificial Intelligence, Vancouver, Canada, August.

Emerson, C. and Ham, I. (1982) An automated coding and process planning system using a DEC PDP-10. *Computer and Industrial Engineering*, **6**(2).

Eversheim, W. and Schulz, J. (1985) CIRP technical reports, survey of computer aided process planning systems. *Annals of the CIRP*, **34**(2).

Ferreira, P.M., Lu, S.C.-Y. and Zhu, X. (1990) A Conceptual Model of Automated Process Planning for the Machined Parts Domain. Proceedings of the 22nd CIRP International Seminar on Manufacturing Systems, University of Twente, Enschede, Netherlands.

Ham, I. and Lu, C.-Y. (1988) Computer-aided process planning: the present and the future. *Annals of the CIRP*, **37**(2).

Houten, F.J.A.M., Erve, A.H., Jonkers, F.J.C.M. and Kals, H.J.K. (1989) PART, A CAPP System with a Flexible Architecture. Proceedings of CIRP International Workshop on Computer Aided Process Planning, Hanover University, Germany, September 21–22.

Houtzeel, A. (1976) The MICLASS System. Proceedings of CAM-I's Executive Seminar-Coding, Classification, and Group Technology for Automated Planning, p-76-ppp-01, CAM-I Inc., Arlington, Texas.

Houtzeel, A. (1980) Computer Assisted Process Planning: A First Step Towards Integration. The CASA/SME Autofact West Conference, November.

Inui, M., Suzuki, H., Kimura, F. and Sata, T. (1987) Extending Process Planning Capabilities with Dynamic Manipulation of Product Models. Proceedings of the 19th CIRP International Seminar on Manufacturing Systems, Pennsylvania State University, USA, June 1–2.

Iwata, K. and Fukuda, Y. (1987) KAPPS: Knowledge and Knowhow Assisted Production Planning System in the Machining Shop. Proceedings of CIRP International Seminar on Manufacturing Systems, Pennsylvania State University, USA, June 1–2.

Iwata, K. and Fukuda, Y. (1989) A New Proposal of Dynamic Process Planning in Machine Shop. Proceedings of CIRP International Workshop on Computer Aided Process Planning, Hanover University, Germany, September 21–22.

Jiang W. and Xu, H. (1987) CAPP Systems and Application in China. Proceedings of the 19th CIRP International Seminar on Manufacturing Systems, Pennsylvania State University, USA, June 1–2.

Lenau, T. and Alting, L. (1986) XPLAN – An Expert Process Planning System. The 2nd International Expert Systems Conference, London, 30 September 1–2 October.

Link, C.H. (1976) CAPP, CAM-I Automated Process Planning System. Proceedings of the 1976 NC Conference, CAM-I Inc., Arlington, Texas.

Lu, S.C.-Y. and Zhang, G. (1990) A combined inductive learning and experimental design approach to manufacturing operation planning. *Journal of Manufacturing Systems*, **9**(2), pp. 103–15.

Lyons, J.W. (1986) The Role of Process Planning in Computer Integrated Manufacturing. Proceedings of the 7th International Conference on the Computer as a Design Tool, London, 2–5 September.

Milacic, V.R., Urosevic, M., Veljovic, A., Miler, A. and Race, I. (1987) SAPT – Expert System Based on Hybrid Concept of Group Technology. Proceedings of the 19th CIRP International Seminar on Manufacturing Systems, Pennsylvania State University, USA, June 1–2.

Ohr, S. (1990) Speeding process planning for the Navy. *Manufacturing Systems*, August.

Schenk, D.E. (1966) Feasibility of Automated Process Planning. Ph.D. Thesis, Purdue University, West Lafayette, Indiana.

Spur, G., Krause, F.C. and Kempt, M. (1990) Advanced Software Tools for the Development of CAPP – Systems. Proceedings of the 22nd CIRP International Seminar on Manufacturing Systems, University of Twente, Enschede, Netherlands.

Steudel, H.J. (1984) Computer-aided process planning: past, present and future. *International Journal of Production Research*, **22**(2).

Tulkoff, J. (1987) Process planning in the computer-integrated factory. *CIM Review*, Fall.

Tönshoff, H.K., Beckendroff, U. and Anders, N. (1989) FLEXPLAN – A Concept for Intelligent Process Planning and Scheduling. Proceedings of CIRP International Workshop on Computer Aided Process Planning, Hanover University, Germany, September 21–22.

Waldman, H. (1983) At one of the country's major helicopter manufacturers. *CAPP*, edited by Tulkoff, J., 1985.

Weill, R., Spur, G. and Eversheim, W. (1982) Survey of computer-aided process planning systems. *Annals of the CIRP*, **31**(2).

Wright, A.J., Darbyshire, I.L., Park, M.W. and Davies, B.J. (1987) EXCAP and ICAPP: Integrated Knowledge-based Systems for Process Planning Components. Proceedings of the 19th CIRP International Seminar on Manufacturing Systems, Pennsylvania State University, USA, June 1–2.

Wysk, R.A. (1977) An Automated Process Planning and Selection Program: APPAS. Ph.D. Thesis, Purdue University, West Lafayette, Indiana.

Wysk, R.A., Chang, T.C. and Ham I. (1985) Automated Process Planning Systems: An Overview of Ten Years of Activities. The 1st CIRP Working Seminar, Paris, 22–23 January.

CHAPTER 7

Implementation of CAPP systems

7.1 INTRODUCTION

The fundamental concepts of CAPP have been defined and discussed in Chapter 6. Implementation of CAPP is a complex task which involves a variety of boundary technologies, especially for new computer techniques (both hardware and software, including the application of AI). In this chapter, some commonly used technology components involved in the implementation of CAPP will be discussed. The format of decision rules will be discussed in section 7.2. In this section the two most common approaches to decision rules – Decision Tables and Decision Trees – will be discussed in detail. The application of Group Technology (GT) will be discussed in section 7.3 and the strategy of planning, defined by forward/backward planning approaches, will be discussed in section 7.4. Following this, the format of input and output will be described. Two alternative input formats, interactive and interface, will be emphasized. The programming languages which have been involved in implementing CAPP systems will be generally discussed. Two commonly used programming languages for implementation of expert process planning systems, PROLOG and LISP, will be commented upon. The expert system (ES) environments, including ES shells and ES toolkits, will finally be discussed at the end of this chapter.

7.2 FORMAT OF DECISION RULES

As Figure 2.2 illustrates, the current approaches of process planning can be divided into two categories, manual and computerized. Here two strategies, traditional and workbook approaches, have been addressed by manual process planning. Computerized process planning can be classified into two variants, generative and semi-generative approaches, which we discussed in the previous chapter. Logical decision is a traditional implementation technique used in computer aided process planning. This is the description of the specifications of the several activities associated with an input and the sequences that are to be followed by the machine. This method has been in use for a long time to help logical decision making. After the increase in the application of computers and the advent of artificial intelligence systems, this has gained popularity. To date, about five alternative formats for decision rules have been recognized in the implementation of CAPP systems; they are (Allen, 1987):

- Decision tables
- Decision trees/decision tables
- Axiomatic
- Rule-based decision trees
- Constraint-based

Among these five decision making approaches, two are commonly addressed in existing CAPP systems, namely decision tables and rule-based decision trees.

7.2.1   Decision tables

Decision tables are tables that are divided by columns and rows. For example, Figure 7.1 shows a decision table for tooling selection. This table corresponds to a specific machine – CORTINI H105, a CNC lathe. The first column of the table represents a four-figure alphanumeric code to indicate the cutting subjects. The first row of the table is the raw material often used in the company. If the specific cutting conditions and cutting material are given, a six-figure alphanumeric code which leads to a specific cutting tool will be generated from the table. Decision tables are often used in process planning. There are several factors that are to be considered while developing a decision table, such as the accuracy, repetitiveness, consistency, size and completeness of the table, to ensure that the table helps effective decision making. The size of the decision table is important. If a decision table is too large, for example several pages of printing, it is difficult for a human to read and interpret. The discipline may fit a menu-driven interactive computer program, although this will not only require excessive memory, but will also reduce the efficiency of decision making.

The table should contain the actual rules and conditions specified in the design. According to the rule representation, decision tables can be classified as follows:

Limited entry decision tables that represent the exact conditions (input values) as true or false entries.

Aluminium

	Alloy steel	Aluminium alloy	Carbon steel	Composite/ Plastic	Copper/ Bronze	Gray cast iron	Stainless steel	Mostly often used raw materials
CF92								
FL11								
FS12								
FT52								
RS11			TC90C S			TC90G C		
. . . .								
XX00	XX00XX	XX00XX	XX00XX	XX00XX	XX00XX	XX00XX	XX00XX	. . . .

CODES I XX00: Four-alphanumeric-code to indicate the cutting subjects
— Codes of cutting dimensions group
— Codes of cutting surface shape
— Codes of process activities

CODES II XX00XX: Six-alphanumeric-code to indicate the tool conditions
— Codes of workpiece materials
— Codes of tools main geometry
— Codes of tool materials

Figure 7.1 An example of a decision table.

Extended entry decision tables that specify the condition but not the value.

Mixed entry decision tables whereby sequenced and unsequenced actions can be entered. Sequenced actions rate a sequence number while unsequenced actions do not rate one.

7.2.2 Rule-based decision trees

Decision trees resemble a graph with a root, nodes, and branches. They are used to represent the results of actions. The root is the source of the tree, and each tree can have only one single root. However a tree can call another sub-tree, for example DCLASS, a typical visible tree structure system. A DCLASS sub-tree can call up to 25 additional sub-trees. When applied to decision making, the branches carry values/ expressions that can be likened to an IF statement, while the branches in series can be likened to an AND statement. Branches can have only two values – true or false. If a branch is true, then it can be passed to the next node. Nodes can be classified into excursive and non-mutually excursive (Allen, 1987). A non-mutually excursive node allows all its successive branches to be true.

Decision trees can either be used as computer codes or represented as data. As a computer code, the tree is converted to a flow chart. The starting node is a root, and every branch represents a decision statement which is either false or true. For a true condition, an action is taken at the corresponding junction, and for a false condition, it is branched out to others or simply terminated. Decision statements can either be mathematical expressions or predicates.

Decision trees have certain definite benefits over decision tables: first, trees can be updated and maintained more easily than can decision tables; second, selected branches of the decision tree may be extended to a considerable depth if necessary, while other branches may be quite short, which is more difficult to do with decision tables; third, some branches of the decision tree may be used to define TYPE and others ATTRIBUTES, which results in relatively small trees; and fourth, trees are easy to customize, visualize, develop, and debug. There are several types of trees which may be developed to aid in classification characterization, selection, and complex decision-making. Figure 11.12 illustrates an example of a decision tree.

7.3   USING GROUP TECHNOLOGY

The fundamental concepts of GT have been discussed in Chapter 4. In terms of process planning, the strategies of using GT are associated with approaches to process planning; for example, the variant, the generative, or the semi-generative approaches.

In the early stage of the development of computer aided process planning, group technology is an essential. All the early CAPP systems are GT based systems. In these systems, the product parts are grouped into different part families. The parts within a part family possess similarities. A master part can usually represent all similarities in the family. The original process plan is made manually and stored in a computer database for the master part. The corresponding process plan is called the master plan. Whenever a process plan is required to be made for a new part, the GT codes will be recognized according to the geometrical attributes and the corresponding part family will be selected. Once the part family is recognized, the master plan will be retrieved. Usually the master plan has a major similarity with the part that needs to be planned. A professional planner is required to modify the master plan according to the specific attributes of the part. The final modified process plan will be printed out as a plan sheet. These early systems are typical variant process planning systems. Even now, most existing process planning systems in the real manufacturing environment are still based on GT. These systems use the concepts of classification and coding, the only difference being that the alpha-numerical systems are constructed in different ways and have different meanings. Many advantages and benefits of using GT on variant approach systems have been discussed in Chapter 4. The biggest disadvantage is that the coding system cannot reflect a direct perception of geometrical shape. Furthermore, the input of code is usually done by human interaction. This requires a skilled person who is familiar with the coding system. Despite this, the manual coding input is a big barrier to the realization of Integrated Process Planning (IPP).

In general, it is difficult to use the GT concept for either a generative or an expert process planning system. This is due to the fact that in such a system decision making usually depends on either specific decision rules input into the rule-base or a specific geometrical feature provided with geometrical modeling.

In some sense, it may be difficult to make a process plan automatically for every part within a part family if it depends only on the recognition of the part family. On the other hand, a process plan which depends on specific decision rules may go too far in terms of facility requirements for manufacture in the workshop. In other words, this plan may be possible in theory, but it is not economical or even possible in practical real production environments. However, tremendous efforts have been made to develop generative-type process planning by different interested groups, using suitable generative coding systems – geometric modelling, expert systems etc. – along with appropriate optimization logic. For example, a group technology primary empirical approach could be enhanced by introducing theory for geometrical and manufacturing pattern recognition as well as manufacturing logic. By starting with a group of parts, it is possible to generate a complete set of parts which belong to the same group. The SAPT system is developed on this concept. SAPT uses a hybrid concept of Group Technology (GT) and Type Technology (TT) (Milacic *et al.*, 1987). The architecture of SAPT is illustrated in Figure 7.12. It is probably the first expert process planning system which involves GT in the knowledge based system. The SAPT system will be further discussed in section 9.3.14.

To use the GT concept in a generative process planning system, the authors have considered another idea and employed it in the development of XPLAN-R, an expert process planning system for rotational parts. A hybrid concept of fuzzy boundary part families and specific decision rules has been employed in XPLAN-R. First, a pattern recognition mechanism established in the parts specification stage distinguishes the fuzzy boundary part families for a geometrical model input by either an interactive or interface scheme. This recognition depends on the size of the part, the shape of the part and the features of the part. After recognition, a rule-decision mechanism is employed to carry out the plan corresponding to each specific part which has been recognized in the part family. At the moment, three fuzzy boundary part families have been introduced into the system. Details of the application of this concept can be found in section 11.7.1.

7.4   FORWARD AND BACKWARD PLANNING

Planning has been defined by Ham and Lu (1989) as the activity of devising means to achieve desired goals under given constraints and with limited resources. In general, the three basic components of any planning activity are goals, constraints, and resources. An intelligent planner, whether a human or a computer program, should have the ability to understand, represent, and manage these three components. Sometimes the term planning is extended to include activities related to plan monitoring, which ensures that plans generated can be executed properly.

To develop a CAPP system, an implementation technique, either a forward or backward planning approach, has to be chosen based on the goals, resources, and constraints. The CAPP systems can take either an initial-to-final or a final-to-initial path to represent the sequences of processes. This means that the system is designed on either a bottom-up or a top-down basis. In some literature, the terms bottom-up and top-down are referred to as forward-planning and backward-planning, respectively. In this book, the terms forward-planning and backward-planning are used.

The forward-planning approach

The forward-planning approach, as the name suggests, develops the system by means of tracing the task of process planning from the raw material to the finished part. This is the actual way that things are done; that is, the workpiece starts with the raw material, various machining processes are carried out, and the final part is obtained. This is a conventional computer aided process planning method oriented toward the variant process planning system. The basic starting point in forward-planning is to simply use the computer as an arithmetic calculator. A natural step from forward-planning is to develop a series of software tools based on clearly understood formulae, which, as a result of minimum operator programming and interaction, will produce the necessary numerical data normally found on the process route or operation sheets issued to the shop floor. This approach was quite widely used in the early stage of the development of CAPP systems. The advantages of this approach are that it is easy to implement a CAPP system and it can easily be used. This approach is inexpensive in terms of investment. The disadvantages of this approach are that it utilizes lower level automation in generating process plans and is difficult to integrate with other computer aided manufacturing systems. The most disadvantageous aspect of the approach is that the popularly used methods, feature based design and feature based process planning, cannot be directly used by means of forward-planning.

The backward-planning approach

The backward-planning approach, contrary to the forward-planning approach, develops the CAPP system by tracing the task of process planning from the finished parts to the raw materials. This is an automated computer aided process planning method oriented toward the generative process planning system. In the backward-planning approach, the computerized system works in the sequence determination of overall strategy, analysis of detail tracks, analysis of the production process for each individual component, etc., as previously listed (the ordered steps of process planning). Backward-planning requires that, in the first instance, the general rules or techniques of the manufacturing strategy must be built into algorithms which can operate on brief input data describing the geometric features and engineering requirements as they relate to individual components. To develop a backward-planning system is usually more expensive than to develop a forward-planning system. The system is more complex and the capacity capability of the system is usually larger than that of a forward-planning system. With the rapid development of feature based design and feature based process planning techniques, the backward-planning approach has attracted more and more attention. Many artificial intelligence based systems are implemented by means of backward-planning. Many integrated process planning systems are also based on the backward-planning approach.

However, since the two methods are not isolated from each other, sometimes a good mixture of forward-planning and backward-planning methods can produce good results in enabling inference manufacturing logic to be built into a CAPP system. In particular, with the recent rapid development of feature based techniques, we may see that forward-planning will also be used and implemented in AI based generative process planning systems.

7.5 FORMAT OF INPUT AND OUTPUT

The input to the process planning system is design data. Generally, the format of input to CAPP systems can be divided into two categories: either a text input or a graphic input. Text input is also referred to as interactive input, where a number of questions in an English or English-like dialogue or a series of alpha-numerical codes are entered through the keyboard of the computer. Graphic input is also referred to as interface input, where the data of the part's geometrical model is gathered from a CAD system.

Most of the variant systems adopt interactive input, especially for coding systems. As discussed in section 4.3, a code is built up through classification and coding by answering a number of predefined questions. These codes are often used to identify the part's families and are associated with standard master plans. The standard master plans are retrieved and edited for the parts. An example of such a code has been shown in Figure 4.2.

So far, both interactive and interface input have been employed in generative systems. If a coding system is used in a generative system, the codes are usually more detailed, and sometimes mix code digits with explicitly defined parameter values. Since a code is concise, it is easy to manipulate. When process capabilities are represented by a code, a simple search through the process capability to match the component code will return the desired process. If an English or English-like dialogue approach is used in a generative system, a specially designed part description language is employed to provide detailed information for process planning systems. A language can be designed to provide all of the information required for the necessary functions of a process planning system. The format can be designed such that functions can easily accomplish their task from the information provided. As an example, an interactive input system has been introduced by the authors (Zhang and Alting, 1989). The input data necessary for the system XPLAN-R is entered into the system by means of user-friendly and highly interactive sessions. The system is menu driven and allows the user to select between single or multiple choice menus, or by inputting variable values.

The format of interface input is mostly used in generative systems. Since a design can be modeled effectively in a CAD system, using a CAD model as input to a process planning system can eliminate the human effort of translating a design into a code or other descriptive form. The increased use of CAD in industry further points to the benefits of using CAD models as process planning data input. In spite of this, the format of interface input is still a fairly undeveloped area due to its complexity. Since it is necessary to enable integrated manufacturing systems, much effort to interface CAPP with CAD has occurred. The terms feature recognition, feature extraction, feature refinement, and geometry reasoning have been used to denote the study discussed in section 8.5. According to a recent investigation, many good approaches for feature based design have been carried out (Joshi and Chang, 1990). These approaches can provide an explicit product model in terms of feature dimensions and feature relationships. These feature based approaches have improved the current CAPP techniques into a new level. However, as we have discussed in Chapter 6, computer aided process planning is a very complex task. It requires not only complete information for a geometric model, but also detailed information for manufacturing. Existing CAD systems cannot provide sufficient manufacturing information for process planning, for instance, tolerances, surface treatment, hardness requirements and so on. If we take into account the integration with job shop scheduling, then the input requirement will have to include some information about the shop floor situation and production control information, which are usually required from a production management database. This particular issue of input information from the shop floor for integrating the process planning function with the production scheduling function will be discussed in section 10.6.3.

The format of the output of CAPP systems can be mainly divided into two terms, either text output, where the process plan carried out by the CAPP system will be directly printed out from a printer or screen, or data output, where the process plan carried out by the CAPP system will be saved in a program which can be retrieved by computer aided manufacturing systems. The text print out (often called the plan sheet or route sheet) usually contains route, processes, operation parameters, machine and tools selected, and sometimes time and cost calculations. It is usually well edited by a word processing program or a built-in output format algorithm. In some cases, a bill of material, a machine sheet, a tooling and fixturing sheet, and a time and cost estimation report can be printed out separately from the system. Generally, the data output is associated with the NC program which is often required by a CAM system. The process plan, which is carried out by the CAPP system, is stored in a program or sometimes directly transferred to a NC path or some other CAM system if the interface has been established between CAPP and CAM.

In terms of integrating process planning with production scheduling, the output of CAPP should also concern the manufacturing resource planning (MRP II). To integrate these functions, the output of CAPP should also be sent to the scheduling module of MRP II to provide the actual status of work in progress on a real-time basis. This is virtually always done by means of a computer terminal directly networked to the scheduling database. Further discussion of integrating process planning and production scheduling will be provided in section 10.6.

7.6 APPLICATION OF ARTIFICIAL INTELLIGENCE IN PROCESS PLANNING

Traditional computer based methodologies are unable to deal with the challenges of fully automated process planning because, although the traditional computer based methodologies may be good at processing data for information-intensive domains, they are not well suited for automatic inference for the knowledge-intensive domain, which automated process planning requires. Rather than simply processing information and data, artificial intelligence (AI) based techniques are designed for capturing, representing, and utilizing knowledge on computers, and hence intelligent manufacturing is certain to play an important role in manufacturing industry. The application of AI or expert systems to process planning has given some promising results. In spite of the fact that the results are still very limited, they are sufficient to stimulate further research. At the present, though, the limited success of expert systems has proved that process planning is a proper field for application of AI. With the advent of expert systems, the knowledge of a process planner can be transferred to the planning system, making it capable of intelligent reasoning, and thereby facilitating the reasoning process. Expert process planning systems can reason intelligently as they organize knowledge at three separate levels: (1) facts, (2) rules, and (3) control strategy (Chang, 1990). The typical expert process planning system would consist of a database that contains the part geometry and the production rules. The production rules perform the transactions in the database to obtain the desired component and decide which rule has to be applied. In what sequence is a matter for the procedural rules of the system, and it is the operation of these that implements the control strategy of the expert process planning system. However, some AI techniques still need further development. Existing expert systems lack adequate mathematical calculation functions. When calculation tasks have to be performed, the expert system usually takes more time than a normal computer program. This disadvantage not only requires more computing time, but also increases the cost. There are also some other problems, since most of the knowledge representation inference engines of the current expert systems are more system designer oriented than process planner oriented. This, in a sense, is the reason why only a few expert process planning systems have been utilized in real production environments. Nevertheless, few will doubt that AI technologies will be developed to improve process planning systems. Emphasis will be given to the development of more user-friendly software products. In addition, a new generation of intelligent systems – learning systems – will emerge. Such systems will respond to the need for continuous re-teaching, with the capability of monitoring actual production experiences and feeding back information to the planning system. The systems can be used for self teaching and training for novices. Further systems are also predicted by Zdeblick, namely distributed planning systems (Zdeblick, 1987). The distributive planning system will tend to take over the character of the original manual planning system with intelligent software and computer systems replacing human skills, knowledge and experience. The common knowledge base will tend to be segmented and individual knowledge bases will be developed at each level covering each area (factory, cell or workstation) in the manufacturing hierarchy.

The detailed technical problems of current expert process planning system technology will be referred to in section 8.5.

7.7 PROGRAMMING LANGUAGES

In a sense, development of CAPP systems has involved almost all of the popular computer programming languages. These languages can be generally divided into two groups: either problem-oriented languages, such as FORTRAN, PASCAL, C, even BASIC, and COBOL, or symbol-manipulation languages, such as LISP, PROLOG, POPLOG, etc.

7.7.1 Problem-oriented languages

In general, problem-oriented languages are designed to solve particular classes of problems. BASIC, as a simple interactive programming language, was used quite widely at the beginning of CAPP development. For two reasons, lack of logic decision ability and emergence of high level languages, BASIC is no longer popular for use in the CAPP area. While BASIC is declining, FORTRAN, also an early computer programming language, is maintaining its superior position in CAPP. FORTRAN has convenient features for performing algebraic calculations and is most applicable to scientific, mathematical, and statistical problem areas. PASCAL, as well as C with the UNIX operating system, are now important in CAPP because of their clear control structure and powerful data structure. Since the task of process planning involves a large number of logical decisions, especially for generative systems, some disadvantages are obvious in the systems implemented by problem-oriented programming languages. Either systems are too large to be evaluated from the economical point of view, such as the task of manpower and executing time consumption, or the capacity and capability of the systems are too limited to deal with problems from real production environments. In some sense, there are also problems from the point of view of portability. This means that the system may be suitable for only very limited and specific tasks. In spite of some disadvantages, the problem-oriented programming languages, such as FORTRAN and PASCAL, are still very popular for the implementation of process planning systems. In particular, the C language is becoming increasingly important here because of its very good integration into the UNIX environment, which has become the standard operating system for many AI workstations. Recent progress of the C++ language has enabled software developers to implement an object-oriented system by a combination of C/C++ language. In the meantime, such problem-oriented programming languages have been used to implement some expert systems, building environments which can overcome the limitation drawbacks. In this vein, systems with specific grammar syntax and format (called expert system shells or toolkits) will be discussed in section 7.8.

7.7.2　Symbol-manipulation languages

In this book, symbol-manipulation languages are emphasized for designing artificial intelligent applications or even Knowledge Based Expert Systems (KBES). Since the characteristic features of ES are symbol manipulation, formal reasoning, recursion and similar activities, two particularly popular standard symbol-manipulation programming languages have been used to date for the development of complete expert process planning systems. They are LISP and PROLOG.

LISP, designed in the 1950s at MIT, stands for LISt Programming language (Winston and Horn, 1981). LISP is a list-processing language, and the fundamental structure is the list. A list is made up of a combination of abstract symbols. As a functional programming language, LISP offers flexibility in writing rules, so that ES builders can specify their own framework of rules. In the last three decades, many LISP dialects have been created in applications throughout the world, the two most commonly used being INTER-LISP and MACLISP. As a matter of fact, there is not much difference in technical superiority between them; the choice between them is only a matter of personal preference or availability. The present trend is towards the adoption of a standard version, COMMON-LISP.

PROLOG, designed in Marseilles in 1972 and taken up in Edinburgh, stands for PROgramming LOGic language (Roussel, 1975). PROLOG is a logic language based on first-order predicate calculus. Programming in PROLOG involves writing logical formulae. These logical formulae indicate logic relations in problems. In PROLOG, ES builders can specify their problem by presenting only the internal relations involved in the problem. Decision rules are written explicitly in formal logic. PROLOG has a built-in backward chaining mechanism which makes it more convenient, but caution needs to be exercised while using it.

In the early stage of development of ES, there was a statement made that LISP was preferred in America while PROLOG was more popular in Europe (Jensen, 1986). This conclusion is uncertain now. From the information collected, it can be seen that both LISP and PROLOG are widely used for the implementation of ES throughout the world. Among the existing EPPSs which are considered in this book, LISP and PROLOG take exactly equally dominant positions, both 30%. It seems that neither language is more superior than the other. Both of them have their own advantages and disadvantages, but both are inconvenient from the viewpoint of mathematical computation. It would seem sensible to design software systems in which the two languages were integrated in such a way that advantage could be taken of the logic features of one (PROLOG) and functional features of the other (LISP). Many such systems have been designed, such as POPLOG, SDL, LOGLISP, etc. (Bonnet *et al.*, 1988).

POPLOG enables ES builders to combine PROLOG, LISP and POP II within one integrated environment. Also, software written in any compiled language, such as C, ADA, PASCAL and FORTRAN, can be linked into POPLOG programs and can be run on standard conventional hardware. The new version of EXCAP developed at UMIST employs POPLOG (Davies *et al.*, 1986).

7.8 EXPERT SYSTEM BUILDING ENVIRONMENTS

Development of expert systems can start from scratch. In this way a programming language, either problem-oriented such as FORTRAN or PASCAL, or symbol-manipulation such as LISP or PROLOG, can be used. Employing a programming language may provide flexibility to tailor the system to the needs of the problem domain, such as organizing the architecture of the knowledge base, selecting the approach for knowledge representation, or choosing the strategy of the inference engine such as forward or backward inference, which have been discussed in section 7.6. However, this will involve considerable effort and manpower.

In comparison, the other method is to base the development of expert systems on ES building environments. The expert system building environments are generally called either expert system shells or toolkits.

7.8.1 Expert system shells

ES shells, also referred to as skeletal systems, are simply stripped-down expert systems, derived from existing expert systems by removing the knowledge base and leaving only the inference engine and support facilities. In some senses ES shells are also called empty systems. A well-known ES shell is EMYCIN-Empty or Essential MYCIN

(Melle *et al.*, 1981). Some other examples are KAS, EXPERT, S1, etc. (Zhang and Alting, 1989).

ES shells have helped to make expert systems more available, extending their application to a wide variety of uses. However, shells have some limitations and drawbacks. For example, it may be difficult to update the rule base, and it may soon become apparent that the system is not able to respond quickly if the knowledge base is increased. It is essential that the knowledge is well represented within the shell and that the user interface has a good explanatory capability.

7.8.2   Expert system toolkits

ES toolkits, also called general-purpose knowledge engineering languages or software toolkits, are very high-level languages or other software systems, designed specially for knowledge engineering applications but capable of being used in other fields. Some of these are based on knowledge representation by production rules, using propositional or predicate calculus; for example, OPS-5 (Forgy and McDermott, 1977), KEE (Intellicorp, 1984), KES (KES PS, 1987), GOLDWORKS (Gold Hill TM, 1987), DCLASS (CAM Software Inc., 1985), etc. ES toolkits can handle many different problem areas and types. They provide more control over data access and search than an ES shell, but may be more difficult to use. These toolkits vary a great deal in the extent of their generality and flexibility.

Many ES shells and toolkits are offered for running on microcomputers such as IBM PCs and compatibles. These can be put into three categories as follows.

- **Inductive tools**, able to generate the rules of small-scale problems from a set of examples: for example, Expert Ease, TIMM, Rule Master.
- **Relatively simple tools** based on production rules and having limited functionality: for example, Insight and Exsys. These are useful as a help to taking the first steps into expert system methods at little cost.
- **Top-of-the range tools** with performances approaching that of the most powerful computers: for example M1, OPS-5, Personal Consultant, Expert. With the continuing increase in the power of microcomputers, in particular 32-bit architecture and large memories, these tools will soon enable professional systems to be constructed with their help, although one cannot at the moment foresee their application to industrial-scale systems.

Thus a great variety of ES building environments have been produced for supporting the development of expert systems. Badiru has listed about 53 existing ES shells or toolkits and system vendors, with the prices in $US and some simple specifications, (Badiru, 1988). However there is no universal tool that will serve for the development of ES in all possible fields. It is vitally important to choose the correct ES building environments because the environments' capabilities must match as closely as possible the user's requirements. The prospective user must be sure that the shell suits the application in mind.

7.9   THE ARCHITECTURES OF CAPP SYSTEMS

One of the most fundamental elements of complex system design is the architecture upon which the system is built. Architecture basically describes the system components, their interfaces, and their relationship to one another. The most important function of a system architecture is that the information flow within the system can be represented explicitly. The details of the architectures may be different from system to system, the information input and output may be based on different schemes, but every automated process planning system must possess five fundamental elements: a knowledge base which is used to store the production rules, an inference engine which is used for knowledge acquisition and representation, a control mechanism which is used for controlling the components communication, a user interface scheme which is used to communicate with users, and a database which is used to store all necessary information used for making process plans.

Figure 7.2 is an example of an architecture of an expert process planning system. The system consists of eight elements. The central element is a knowledge base. The knowledge base may be implemented in many different ways. In this particular example, the knowledge base is implemented by a DCLASS tree processor, so all the production rules are implemented in the form of a tree structure. The detailed description of the working function of this system will be found in Chapter 11. In continuing, we are going to introduce a framework for a conceptual model of process planning. As shown in Figure 7.3, this model consists of five modular activities, not only for process planning alone, but also for the integration of process planning with shop floor control. The following is a concise summary of the planning activities within each module as described by Ham and Lu (1989).

Despite these difficulties, searching for such a generic conceptual framework is very critical and has been pursued by many research organizations. A good example of such a framework is proposed in (CAM-I, 1988). Figure 7.3 shows the structure of the proposed framework, which consists of five modular activities. It is interesting to note, as pointed out by the proposer, that process planners do not always plan in a breadth-first manner within this hierarchical structure. The

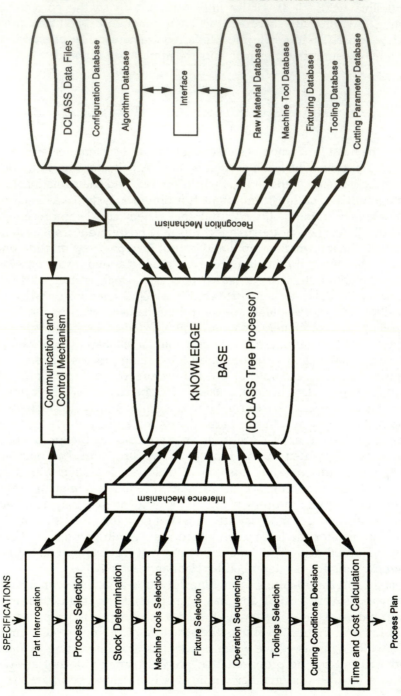

Figure 7.2 An architecture of an expert process planning system.

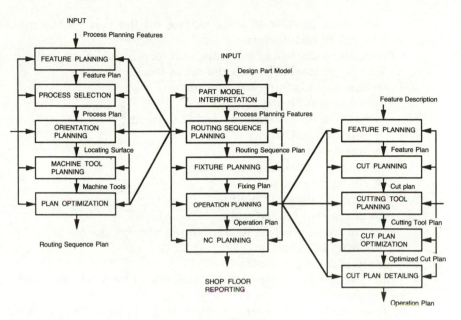

Figure 7.3 A conceptual framework for CAPP systems (CAM-I, 1988).

following is a concise summary of the planning activities within each module.

1. **The part model interpretation module:**
   - input: part models from design
   - perform: feature extraction
   - output: manufacturing features that need to be planned for
2. **The routing sequence planning module:**
   - input: manufacturing features that need to be planned for
   - perform: the number and order of operations of the manufacturing workstation for each operation, and the prior and subsequent geometry for each operation
   - output: routing sequence plan
   - this module normally contains five sub-modules:
     - feature planning sub-module
       * organize manufacturing features into a feature-access graph to represent the precedence of the machining of features based on their access
       * decompose composite or complex manufacturing features into simpler features
     - process selection sub-module
       * determine the necessary manufacturing process required to make each feature
     - orientation planning sub-module

* determine possible locating surface on the part for the machining of each feature
* consider feasible tool axis approach for a feature such that the axis is approachable with the set of locating surface
- machine tool planning sub-module
  * to select feasible machine tools
  * a machine tool module is needed to check:
    • relative motions
    • travel limit
    • additional machining attachments
    • desirable accuracy
- plan optimization sub-module
  * to group cuts into common fixing setups
  * group fixing setups into common machining operations
  * ordering of operations

3. **The fixture planning module:**
   * input: part module from design, features to be machined, their locating surfaces
   * perform: determine the correct fixing device to locate the part and restrain it for the machining operations
   * output: fixing devices and methods
   * functional requirements for fixtures are:
     - resting equilibrium
     - deterministic location
     - clamping equilibrium
     - total constraint
     - try to use standard fixtures whenever possible

4. **The operation planning module:**
   * input: part module, feature volumes to be machined, machine tools, and fixtures
   * perform: details specifications related to the execution of a cut
   * output: to determine cutting tools, number and levels of cuts, the order of cuts, the machining parameters for a cut, instructions for a machinist, and NC tape specifications
   * there are also five sub-modules
     - feature planning sub-module
       * decide on a sequence for machinable features
       * decompose machinable features into machinable volumes
     - cut planning sub-module
       * decompose machinable volumes into specific cuts that can be performed on the specified machines
       * select levels of cuts (rough, semi-finish, finished)
     - cutting tool planning sub-module
       * determine the cutting tool to use based on geometry and tolerance definition of each machining cut

- cut plan optimization sub-module
  * group cuts that have intersecting tool lists into the same tool change
  * order the plan into a good machining sequence
- cut plan detailing sub-module
  * selection of tool holder
  * specify cut depth, feed, speed data
  * report the operation plan

5. **NC planning module:**

To implement a generative CAPP system based on the above framework poses many great technical challenges which have been the subjects of study in the USA in recent years. A brief summary of these challenges follows:

- knowledge base representation and organization
  - modular, iterative, and multiple perspective nature of the task
- product definition (should be complete, flexible, comprehensive, exact, etc.)
- feature definition, representation, and recognition
  - primary activities involved in feature recognition:
    * identify machined faces
    * create machined volume(s)
    * classify machined feature(s)
- geometric reasoning
  - reasoning about shapes other than rectangular, prismatic features (can't treat as macros with standard attributes)
- process optimization
  - transform a list of independent cut specifications into an optimal, linearized machining sequence

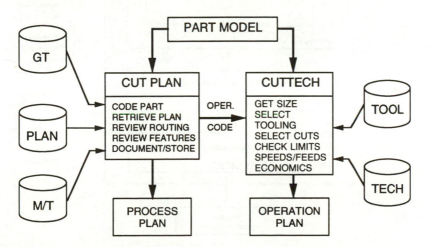

Figure 7.4 CUTTPLAN/CUTTECH system architecture (Zdeblick, 1985).

Many AI-based approaches have been researched to respond to the above challenges. While knowledge-based expert systems are still the most commonly employed technique in developing generative CAPP, other AI-based techniques, such as machine learning, case-based reasoning, qualitative physics, etc., have also been developed recently. We expect that these research efforts will soon result in more enhanced reasoning and representation approaches to intelligent CAPP.

Recently, research results have evidenced that for an integrated process planning system, to implement a knowledge base will require not

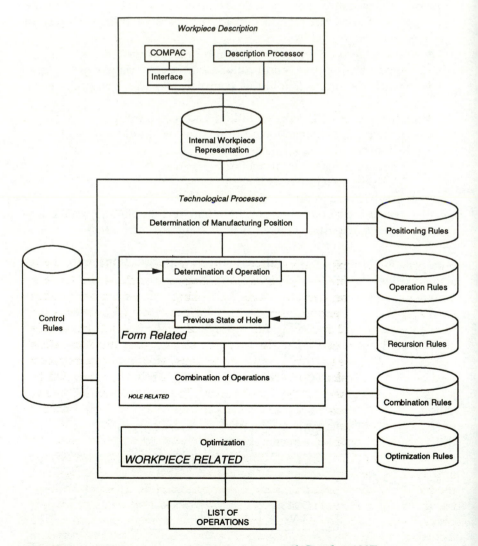

Figure 7.5 DOPS system architecture (Major and Grottke, 1987).

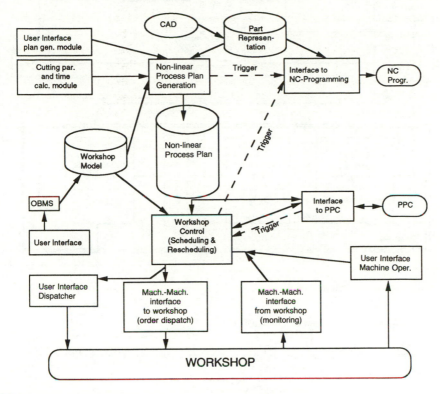

Figure 7.6 FLEXPLAN system architecture (Tönshoff *et al.*, 1989).

only a symbol-manipulation language such as LISP or PROLOG, but also some object-oriented programming language such as the C/C++ language. For some complex decision making, an expert knowledge base system is not enough: some fuzzy logic and neural networks are needed. Figure 7.2 indicates the trends of this. The left-hand side shows, in a series of blocks, the working steps. From the illustration one can see that the system works in nine steps, from part interrogation to time and cost calculation. All these working steps are controlled by a communication and control mechanism. The information communication between each step and the knowledge base will be handled by means of an inference mechanism, as indicated in the figure. At the right-hand side of the figure are two interfaced databases. All geometrical models and shop floor data are stored in the databases, and the data are controlled and retrieved by means of a recognition mechanism. As an example, this system may not be a perfect system, but in Chapter 11 readers will find out that such a system can provide process plans which are good enough to be used in the real manufacturing environment. In Figures 7.4–7.13 several different system architectures are illustrated.

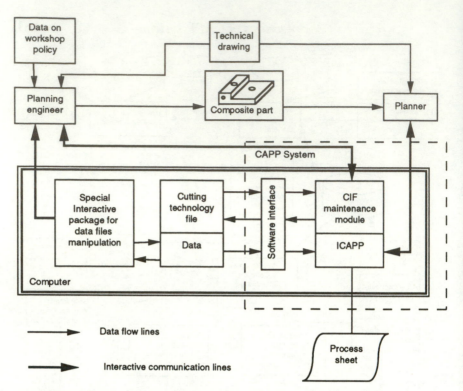

Figure 7.7 ICAPP system architecture (Eskicioglu and Davies, 1983).

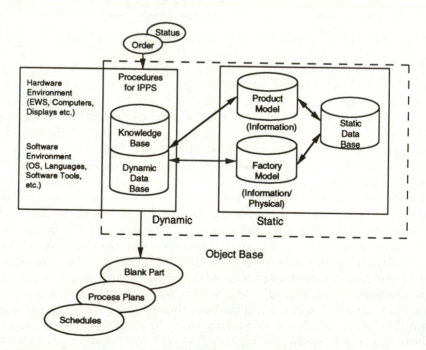

Figure 7.8 IPPS system architecture (Iwata *et al.*, 1990).

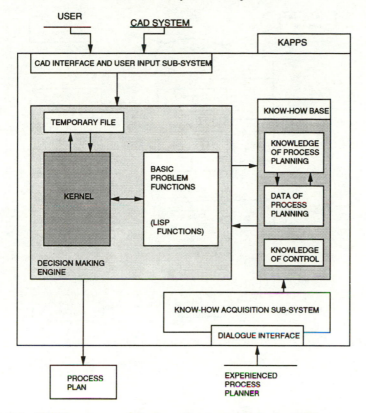

Figure 7.9 KAPP system architecture (Iwata and Fukuda, 1989).

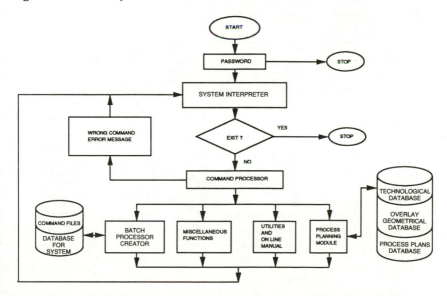

Figure 7.10 PRICAPP system architecture (Pande and Walvekar, 1985).

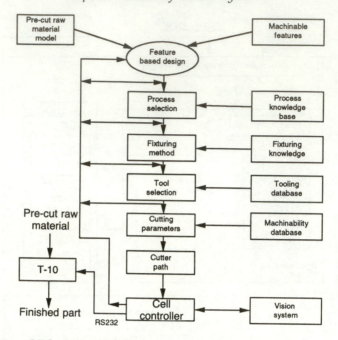

Figure 7.11 QTC system architecture (Kanumury *et al.*, 1988).

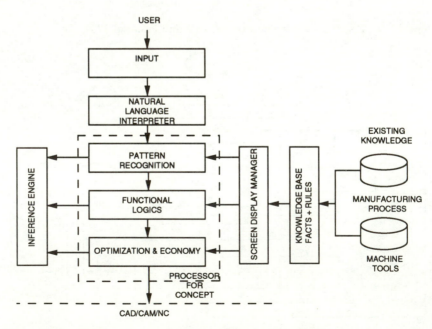

Figure 7.12 SAPT system architecture (Milacic *et al.*, 1988).

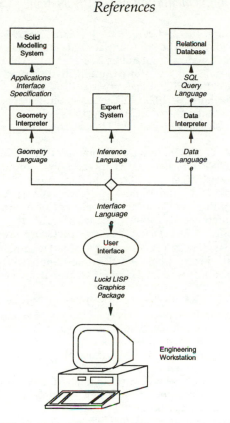

Figure 7.13 XCUT system architecture (Hummel *et al.*, 1989).

REFERENCES

Allen, D.K. (1987) An introduction to computer-aided process planning. *CIM Review*, Fall.

Badiru, A.B. (1988) Expert systems and industrial engineers: a practical guide to a successful partnership. *Computers in Engineering*, **14**(1), pp. 1–13.

Bonnet, A., Haton, J.-P. and Truong-Ngoc, J.-M. (1988) *Expert Systems Principles and Practice*, Prentice Hall International Ltd., UK.

CAM-I (1988) *XPS-E Revisited: A New Architecture and Implementation Approach for Automated Process Planning Systems*. CAM-I DR-88-PP-02.

CAM Software Inc. (1985) *DCLASS Technical and System Manual*, CAM Software Inc., Provo, Utah, USA.

Chang, T.C. (1990) *Expert Process Planning for Manufacturing*, Addison-Wesley Publishing Company, p. 281.

Davies, B.J. (1986) Application of expert systems in process planning. *Annals of the CIRP*, **35**(2).

Davies, B.J., Darbyshire, I.L. and Wright, A.J. (1986) Expert Systems in Process Planning. Proceedings of the 7th International Conference on the Computer as a Design Tool, London, September 2–5.

Eskicioglu, H. and Davies, B.J. (1983) An interactive process planning system for prismatic parts (ICAPP). *Annals of the CIRP*, **32**(1).

Forgy, C. and McDermott, J. (1977) OPS: A domain-independent production system language. *International IJCAI*, **5**, pp. 933–9.

Gold Hill TM (1987) *Goldworks Programmer User's Guide*, Gold Hill TM, Intellicorp Limited.

Ham, I. and Lu, S.C.-Y. (1989) New Development of CAPP in USA and Japan. Proceedings of CIRP International Workshop on Computer Aided Process Planning, Hanover University, Germany, September 21–22.

Hummel, K.E., Brooks, S.L. and Wolf, M.L. (1989) XCUT: An Expert Systems for Generation of Process Planning. Proceedings of International Industrial Engineering Conference, Toronto, Canada, May 12–18.

Intellicorp (1984) *Nucleic Acid Sequence Analysis*, Intellicorp brochure.

Iwata, K. and Fukuda, Y. (1989) A New Proposal of Dynamic Process Planning in Machine Shop. Proceedings of CIRP International Workshop on Computer Aided Process Planning, Hanover University, Germany, September 21–22.

Iwata, K., Fukuda, Y. and Sugimura, N. (1990) A Proposal of Knowledge Based Structure for Integrated Process Planning System. Proceedings of the 22nd CIRP International Seminar on Manufacturing Systems, University of Twente, Enschede, Netherlands.

Jensen, J. (1986) Computer Integrated Manufacturing and Expert System. ATV-Semapp Workshop, Copenhagen, March 6.

Joshi, S. and Chang, T.C. (1990) Feature extraction and feature based design approaches in the development of design interface for process planning. *Journal of Intelligent Manufacturing*, **1**, pp. 1–15.

Kanumury, M., Shah, J. and Chang, T.C. (1988) An Automatic Process Planning System for a Quick Turnaround Cell – An Integrated CAD and CAM System. Proceedings of USA–Japan Symposium of Flexible Automation, ASME, July 18.

KES PS (1987) *Knowledge Author's Manual*, KES PS, Software A&E Inc.

Major, F. and Grottke, W. (1987) Knowledge engineering within integrated process planning systems. *Robotics and Computer-Integrated Manufacturing*, **2**, pp. 209–13.

Marshall, P. (1985) Computer-aided process planning and estimating as part of an integrated CADCAM system. *Computer-Aided Engineering Journal*, October.

Melle, van W., Shortliffe, E.H. and Buchanan, B.G. (1981) EMYCIN: A domain-independent system that aided in constructing knowledge-based consultation programs. *Machine Intelligence*, Infotech State of the Art, Report 9, no. 3.

Milacic, V.R., Urosevic, M., Veljovic, A., Miler, A. and Race, I. (1987) SAPT – Expert System based on Hybrid Concept of Group Technology. 19th CIRP International Seminar on Manufacturing Systems, Pennsylvania State University, USA, June 1–2.

Pande, S.S. and Walvekar, M.G. (1990) PRICAPP: A computer assisted process planning system for prismatic components. *International Journal of Production Research*, **28**(2), pp. 279–92.

Roussel, P. (1975) *PROLOG, Manuel de reference et d'utilisation, Group d'Intelligence Artificielle*. UER de Luming, Aix-Marseilles University, France.

Tönshoff, H.K., Beckendroff, U. and Anders, N. (1989) FLEXPLAN – A Concept for Intelligent Process Planning and Scheduling. Proceedings of CIRP International Workshop on Computer Aided Process Planning, Hanover University, Germany, September 21–22.

Winston, H. and Horn, B.K.P. (1981) *LISP*, Addison Wesley, USA.

Wright, A.J., Darbyshire, I.L., Park, M.W. and Davies, B.J. (1987) EXCAP and ICAPP: Integrated Knowledge-based Systems for Process Planning Components. Proceedings of the 19th International Seminar on Manufacturing Systems, Pennsylvania State University, USA, June 1–2.

Zdeblick, W.J. (1985) CUTPLAN/CUTTECH: A Hybrid Computer-Aided Process and Operation Planning System. Proceedings of the 1st CIRP Working Seminar in CAPP, Paris, January 22–23.

Zdeblick, W.J. (1987) Process Planning Evolution: The Impact of Artificial Intelligence. Proceedings of the 19th CIRP International Seminar on Manufacturing Systems, Pennsylvania State University, USA, June 1–2.

Zhang, H.C. and Alting, L. (1989) Expert Process Planning Systems: The State-of-the-Art. Proceedings of the 21st CIRP International Seminar on Manufacturing Systems, Stockholm, Sweden, June 5–6.

CHAPTER 8

Feature based design and IGES/PDES/STEP

8.1 INTRODUCTION

The traditional design approach has almost been replaced completely by Computer Aided Design (CAD) in most manufacturing facilities today. With the advent of CAD there has been a noticeable change in the productivity and quality of the design generated. CAD has also aided in bridging the gap between design and manufacturing facilities and the storage of data inside the manufacturing facility. If CAD has been the cause of so much success, then what was the need to look further for better design methods? The reason was that, as in most other technologies, there were inherent drawbacks in CAD and it fell short in certain areas that are crucial to survival, especially in today's highly competitive manufacturing arena. These drawbacks have necessitated the need for new methods of incorporating design ideas on paper; and this has led to the birth of feature based design and feature recognition techniques. The traditional design approaches, the inception of CAD, the drawbacks, and the inception of the feature based design approach will be the focus of this chapter. The following sections will discuss the feature based design and concepts of feature recognition.

CAD was the result of the shortcomings felt in the traditional manual design approaches that were in practice in the early era of manufacturing. The need to shorten the design time, increase the accuracy of the drawings produced, and improve the design details (like tolerance, dimensions, and property of materials) were acutely felt, and with the advent of computers it did not take very long to come up with the idea of incorporating computers in the field of design. Computer aided design involves making use of the computer to develop, analyze, and/or modify an engineering design. The evolution of CAD systems has largely been attributed to the developments made in computer graphics. Although interactive computer graphics and interactive graphics exchange systems form only part of the CAD system, this was supposed to be the essential technological foundation for CAD.

CAD systems use four different types of graphic representation schemes: 2D drafting, wire-frame systems, 3D surface, and 3D solid models. Among these, solid models provide the most complete part representation and consequently provide an important piece of the CAD/CAPP link. For this reason the rest of this section will discuss 3D solid modelling, as it is likely that more and more process planning systems will be using this approach as their main source of input.

There are at least six different methods for constructing the 3D object within a solid modelling system. They are: pure primitive instancing, spatial occupancy enumeration, sweeping, constructive solid geometry (CGS), and boundary representation (B-Rep). Of these methods the last two, CGS and B-Rep, are the most significant and will be discussed in detail. The CGS modeler uses Boolean operators and primitive solids for constructing the part. The primitive solids are either defined through explicit surface equations or can be generated from a combination of half spaces. On the other hand, the B-Rep method consists of computing all the bounding edges for all the surfaces.

After an object is constructed, the modeler converts the processed input to a data structure, maintaining the geometry and topology of the part. CGS stores the part as a tree, where the leaves of the tree are the primitives and the nodes are the Boolean operators. B-Rep data structure is different in that it has linked lists of all the vertices, edges, and the faces, and thus incorporates the geometry and topology of the part in the data storage. From research and experience it has been found that although extensive computations are required, B-Rep modelers provide a data structure that is more detailed and more convenient for use by the application systems. The essential difference between CGS and B-Rep storage structure is that, whereas CGS stores the instructions for how to construct the part in the system, B-Rep actually stores the geometry of the part. An advantage of the B-Rep models is that manufacturing information, such as dimensions and tolerances, are also associated with the data structure. On the other hand, for complicated polyhedrons such as holes and cones, additional data elements are needed with B-Rep to store curved surface equations. From the manufacturing point of view, B-Rep has a limitation: it stores the features of the part implicitly, and this necessitates an intermediate step to reinterpret and create manufacturing-specific features. This will later be addressed as one of the major drawbacks of CAD systems.

8.2   IGES/PDES/STEP AND DATA SHARING

Clearly, it is not the physical hardware connections between computers that are the major issue in data sharing; it is incompatible software. The root of the problem is proprietary data representations, that is, vendor-specific data forms. More often than not, the vendors of computer applications store the data which is required and produced by their systems in their own proprietary format.

For example, once the design of the product has been completed on a CAD system, it is stored in a data file. Some of the information in that data file represents the shape and size of the product. In an integrated information systems environment, the designer should be able to send that data file over to the manufacturing planning system. The same data would then be used by the planning system to determine manufacturing processes for the product, based in part on its specified shape and size.

If the planning system can read the contents of the design data file, it can obtain the shape and size information it needs. It might be said that these two applications are integrated. But it is a fact today that, if two commercial products are integrated, it is likely that they were developed by and purchased from the same vendor. Furthermore, it is also likely that they were intentionally designed to work together from their inception. Often, it is the case today that applications offered by the same company are not integrated.

IGES stands for Initial Graphics Exchange Specification, which is a neutral data exchange format for such data sharing purposes. IGES was first published in 1980 and was updated in 1983, 1986, 1988, and 1990 (Reed *et al.*, 1990). IGES was originally developed to provide a means for exchanging engineering drawing data between CAD systems for three-dimensional geometrical models. Its goal is to allow CAD data to be exchanged between systems built by different manufacturers. While it is good for wire-frame and boundary representative models, it is not convenient for representing the solid models which are widely used today and which are the trends for the geometrical models of tomorrow. One more problem with IGES is an outcome of the way CAD vendors implement the software that is required to translate their data to and from the neutral IGES data file. When IGES data is passed between design systems, considerable human interpretation and manipulation of data may be required. Since IGES was designed primarily as a mechanism for file exchanges between CAD systems, it is not able to support shared databases between dissimilar product life cycle applications.

With IGES, the user can exchange product data models in the form of 2D and 3D wire-frame representations as well as surface representations. Translators convert a vendor's proprietary internal database format into the neutral IGES format and from the IGES format into another vendor's internal database format. The IGES effort has not focused on specifying a standardized information model for product data. IGES technology assumes that a person is available on the receiving end to interpret the meaning of the product model data.

In order to solve the shortcoming of IGES, the International Organization for Standardization (ISO) is currently involved in the development of a new international standard for the exchange of information related to automated manufacturing. The development standard is informally called the Standard for The Exchange of Product model data (STEP). STEP is intended to address the issue of product data sharing between different computer applications running on different computer systems within the same or different organizations.

STEP goes beyond IGES both in the breadth of its information content and in the sophistication of its information system methodologies. STEP will provide a standard, neutral format for product data created and shared by different applications. Neutral means that the STEP data format will not favor one particular vendor. In addition, STEP development is including the definition of subsets of product data that are specifically required for particular usage contexts. These subsets are called application protocols.

STEP application protocols address the issues of completeness and unambiguity of data transfer by specifying in advance what data should be transferred in a particular context, thereby alleviating the need for vendors to make problematic assumptions. Application protocols are those parts of STEP that are relevant to a particular data-sharing scenario (Palmer and Gilbert, 1991).

There are four major technical challenges facing the developers of STEP (Carver and Bloom, 1991):

- The exchange of data is different from the exchange of information. Data must be transmitted accurately and without any changes. In contrast, information, although composed of data, must be understood and interpreted by the receiver. Furthermore, the receiver must be able to apply the information correctly in new situations. The first challenge is that STEP is a standard for information, not just data.

- The need for STEP to be extendable to new products, processes, and technologies, requires a more abstract representation of the information than in previous standards. Regardless of their equipment or process, users must be able to obtain the information necessary to do something from the STEP representation of a product. Therefore, the second challenge is that the development of STEP must include the development of an architecture or a framework for the exchange of information, not just a means or format for storing information.

- The wide range of industries and the diversity of product information covered in STEP is beyond that of any previous digital standard. The variety of attributes and parameters, such as geometric shape, mechanical function, materials, assembly information, and date for manufacture, is immense. Also, the industrial base, the number of industries involved, is enormous; even greater is the number of technical disciplines that are involved. Moreover, STEP must be flexible and extensible so that new information and additional application protocols can be added and can be upwardly compatible. Therefore, the third challenge is that the scope and complexity of STEP is far beyond any previous standards effort.

- Traditionally, standardization is a process that devises an approach encompassing a variety of existing vendors' methods and builds the best solution available without penalizing some vendors more than others. In the case of STEP, there is no existing implementation. Thus the fourth challenge: the technology to support STEP must be developed at the same time that the standard is evolving.

The consensus approach to meeting the above challenges is to start with conceptual information models. STEP will consist of a set of clearly and formally defined conceptual models and a physical exchange protocol based on these models. The conceptual models will be combined into a single model with a standard interface to a shared database.

PDES, which stands for Product Data Exchange using STEP, refers to the US activities in support of the development of STEP (Clark and Libes, 1992). In April 1988, several major US technology companies incorporated as PDES Inc., with the specific goal of accelerating the development of STEP in the USA. PDES will help establish a standard digital representation for product data. The specifications already developed by the PDES effort have been submitted to the ISO as a basis for the evolving international standard STEP. As the PDES and STEP efforts share common goals, they are sometimes referred to jointly as PDES/STEP, or simply just as STEP.

PDES/STEP is currently developed by a number of organizations, both nationally and internationally. These organizations include:

- IGES/PDES Organization
- ISO TC184/SC4
- ANSI US Technical Advisory Group
- PDES Inc.
- NIST National PDES Test-bed

8.3   FEATURE DEFINITION

Features are generic shapes with which engineers associate certain attributes and knowledge useful in reasoning about the product. Features encapsulate the engineering significance of portions of the geometry and, as such, are important in product design, product definition, and reasoning for a variety of applications. Many researchers in the past have proposed a lot of definitions for a feature and some of them are listed as follows:

- Codifiable properties derived from a taxonomy of shapes for a particular classification scheme.
- A specific geometric configuration formed on the surface, edge, or corner of a workpiece intended to modify outward appearance or to achieve a given function.
- A characteristic volume that describes a portion of a part such as a hole, boss, pad, pocket, etc.

In order to encompass all the different views of a feature we propose the following definition: A feature is a region of interest in a part model.

Example of features

Consider, as an example, a turned shaft which has a keyway at one end, and the other end of the shaft is knurled. If the keyway is non-standard, then its dimension will be specified. Its location will also be given. Now coming to features, both the keyway and the knurl are considered as features of the part. In the first case the geometry is fully detailed, and in the second case the geometry of the area is defined and an attribute is tagged to it. In what follows, we consider the area to be a feature, the attribute being additional information for that feature, often relevant to the manufacturing process.

8.4 FEATURE CLASSIFICATION

Features can be broadly classified into two main categories, as follows:

Explicit features: All the geometric details of the feature are fully defined.

Implicit features: Sufficient information is supplied to define the feature but the full geometric details have to be calculated when required.

As an example, consider a cylindrical blind hole. An explicit representation in a solid modeler will contain details of the cylindrical surface and the surface at the base of the hole, together with the equation of the edge curves. An implicit representation of the same hole might specify the center line, depth, and radius. The explicit information can be computed from the implicit data.

There are seven general features as follows:

- Circular through hole
- Circular pad

- Rectangular pad
- Rectangular pocket with an interior rectangular boss with a blind hole
- Face (area to be plated)
- A bevelled edge
- A corner break

There are many types of features in application contexts, but this text will deal with the most important ones.

A component is designed to meet certain functional objectives. These dictate not only the overall shape of the part, but also its details. Those features which bring about the functional aspect of the component are referred to as **functional features**. They could be in the form of cooling slots, fixing holes, strengthening webs, and so on. The geometric model from the design stage will be a representation of the part to be manufactured. The designer's functional features are generally incorporated into the representation.

The information provided from design to process planning is a representation of the finished part. It is the process planner's function to specify how the part will be manufactured. The features which define the attributes associated with manufacturing are called the **manufacturing features**. In many cases the manufacturing features are almost similar to the design features. A hollow cylinder (hole) for a process planner is not just a hole; he/she needs further information such as whether it is a bored hole, a reamed hole, and so on.

Features are incorporated in the part model from the beginning. Generic feature definitions are placed in a library from which features are picked up by specifying the dimension and location parameters, as well as various other attributes. As shown in Figure 8.1, once the features are pulled out, various operations can be performed on them such as adding and deleting. Design with features allows a higher level of abstraction for design. This method is very pragmatic in nature because the designer has a set of features to choose from, and he/she builds an artifact using these features which are essentially building blocks for the product definition. The function of the monitor here is to guide the designers in their work. It will point out mistakes made by the designers and help them rectify them. As an example, the monitor will not allow a designer to combine the features in ways which the system cannot interpret.

After the design is completed, it is stored in the primary representation and, for the downstream applications, the design is routed through the secondary representation. This conversion can be done by using the knowledge coded into computer programs. Thus this method allows the development of intelligent CAD systems. Hence the ability of the system to successfully convert the information is the core of the entire intelligent CAD system.

8.5   FEATURE RECOGNITION

Feature recognition in simple terms means extracting and recognizing the features. This is defined as converting a model of lower level entities into a model of higher level entities, or converting a geometric model that consists of lower level entities like lines, points, etc. into a feature based model, which is defined in terms of higher level entities like holes, grooves, pockets, etc. The main reason for this conversion is the ability of the features to be associated with the knowledge about the way the feature can be manufactured. Then, in the feature recognition process, once the geometric model is constructed or created by the designer, the computer software or programs in the system should process the database that defines the design, and automatically extract and recognize the features.

A simple schematic of a feature recognition process is illustrated in Figure 8.1.

To satisfy the needs of feature recognition processes and accomplish the task, the following points are necessary with respect to the feature recognizer:

- We should have a proper representation scheme. This means that the scheme created or used for feature recognition should facilitate automatic feature recognition.
- Each feature should have a unique definition. In the representation scheme that is developed or used for feature recognition, there will be a host of feature definitions that help in recognizing and classifying the features. For proper recognition, no two features should have the same definition, and no feature should have more than one definition.
- The inference mechanism or procedure used should be consistent and complete. In other words, the procedure should be able to identify the features completely and correctly.

There are a number of approaches used to recognize the features both from 2D and 3D CAD representations. Some of these are listed as follows (Joshi and Chang, 1990):

Figure 8.1   Feature recognition process.

- Syntactic pattern recognition
- State transition diagrams and automata
- Decomposition approach
- Expert system/logic
- CSG (set theoretic) approach
- Graph based approach

Since these approaches have been well discussed by Joshi and Chang, we are not going into any details; instead we shall briefly introduce the syntactic pattern recognition and graph based approach. Readers interested in the details of these approaches can refer to Joshi and Chang's relevant publications.

8.5.1   Syntactic pattern recognition

This method is similar to language processing, where a sentence is analyzed to check whether its syntax is correct. A picture is represented by some semantic primitives written in a picture language. A set of grammar consisting of some rewritten rules defines a particular pattern. A parser applies the grammar to the picture. If the syntax of the picture language agrees with the grammar, then the picture is classified as belonging to a particular pattern class. This can be visualized as similar to checking a computer program by a compiler.

An example of this method is shown in Figure 8.2. In the two dimensional language, a hole is represented by its projection, and a pattern primitive, as shown in the figure, is defined to show four directions. The picture is then substituted by the pattern primitives by substituting each line segment with the corresponding primitive. Thus the above picture can be replaced by the pattern BBCCBBAABB. From this a general rule or grammar for holes can be defined as $C^nB^mA^n$. Similarly the grammar is defined for other features. Whenever a picture is to be recognized, it is compared with the grammar after it is converted into the picture language (Chang, 1990).

8.5.2   Graph based approach

In this approach, the graphs and networks are used to facilitate the feature recognition. A graph normally consists of nodes, arcs, and

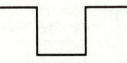

Figure 8.2   A simple hole drawing (Chang, 1990).

attributes. Here, we introduce a graph based approach, namely Attributed Adjacency Graph (AAG) developed by Joshi and Chang (1988).

The AAG is defined as a graph

$$G = \{N, A, T\}$$

where N is the set of nodes,
        A is the set of arcs, and
        T is the set of attributes.

The simple rules in the construction of the AAGs are:

1. For every face f in F, there exists a unique node n in N.
2. For every edge e in E, there exists a unique arc a in A, connecting the nodes n_i and n_j, corresponding to face f_i and face f_j, which share the common edge e.
3. Every arc a in A is assigned an attribute T, where t = 0 if the faces sharing the edge form a concave angle, and t = 1 if the faces sharing the edge form a convex angle.

Figure 8.3 shows an example part and its AAG. The AAG is represented in matrix form in the computer. This can be thought of as similar to the node–arc incident matrix. Each node stores the pointer to the corresponding face information, thus providing the link with the B-Rep also.

Since the features are recognized based on their definition, they need to be precisely defined. For this, each instance of the feature must be identified, and incorrect instances should not be recognized. Thus the definition of features involves determining the minimal set of necessary conditions that classify the feature uniquely.

To illustrate this point, the example of slots shown in Figure 8.4 can be considered. The slots can be defined based on the topological and geometric relations of the faces that form the slot. The difference between the definitions of the three slots is the value of the angle formed by the faces with the base of the slot. The topological information is identical. A generic definition for the class of slots could be based on the following properties:

- F1 is adjacent to F2
- F2 is adjacent to F3
- F1 forms a concave angle with F2
- F3 forms a concave angle with F2

This provides a generalized classification of features and a framework for representing the features in a hierarchical manner. The specific information for further classification can be obtained from the geometric information. One major advantage of this hierarchical organization is the reduction in time required to recognize a feature. A second

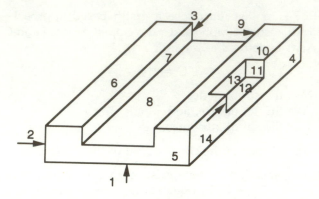

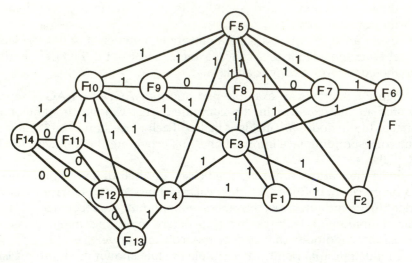

Figure 8.3 An attributed adjacency graph (AAG) (Joshi and Chang, 1988).

advantage comes from the notion of inheritance. As the instance of a particular sub-class is also an inheritance of its super-class, the properties of the super-class can be inherited without explicitly being repeated. Hence this provides a more compact representation.

The properties of the AAG that define a pocket are given below:

- The graph is cyclic
- It has exactly one node n with number of incident 0 arcs equal to the total number of nodes −1
- All the other nodes have degree 3
- The number of 0 arcs is greater than the number of 1 arcs (after deleting the node n)

These properties of the pocket are general enough to identify a wide range of pockets formed. Similar rules are defined for each feature and

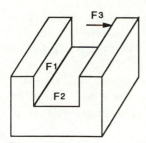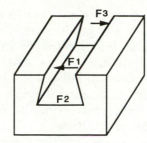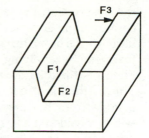

F_1 adjacent to F_2	F_1 adjacent to F_2	F_1 adjacent to F_2
F_2 adjacent to F_3	F_2 adjacent to F_3	F_2 adjacent to F_3
Angle between F_1 and $F_2 = 90$ degrees	Angle between F_3 and $F_2 = 90$ degrees	Angle between F_1 and $F_2 < 90$ degrees
Angle between F_3 and $F_2 < 90$ degrees	Angle between F_1 and $F_2 > 90$ degrees	Angle between F_3 and $F_2 > 90$ degrees
(a)	(b)	(c)

Figure 8.4 Different slot types and their recognition rules (Joshi and Chang, 1988).

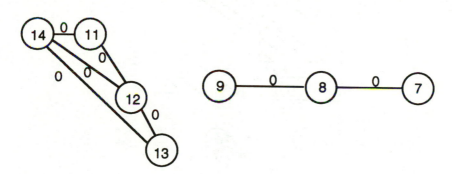

Figure 8.5 Sub-graphs of AAG.

the rules are unique for each feature. In the case of the blind hole, the AAG corresponds to the triangular hole, and additional tests are required to identify the feature.

The features in the part are sub-graphs of the complete AAG and to recognize the features the AAG has to be partitioned into sub-graphs. But this search for the sub-graphs is an isomorphism problem and is computationally exhaustive.

To get around this problem of identifying the parts of a graph that might form features, a heuristic method is proposed. This is based on the observation that a face that is adjacent to all its neighbors with a convex angle does not form part of a feature. This heuristic is used as a basis to partition the graph into sub-graphs that could correspond to features. By applying this heuristic, the AAG of the part shown in Figure 8.3 breaks down into two sub-graphs as shown in Figure 8.5. This method is also known as deleting the nodes. The effect of deleting a node from the graph only corresponds to eliminating a face from being considered as a part of a feature.

The next step in the recognition is analyzing the sub-graphs to determine the feature types. The recognizer incorporates all feature recognition rules and determines the feature type which the sub-graph corresponds to.

Sometimes, while analyzing the sub-graphs, the sub-graphs may not correspond to any of the feature definitions that are incorporated in the feature recognizer. This is because the primitive features may not exist by themselves, but instead form an intersection of features. There are several interactions that are possible, and the two most common types are:

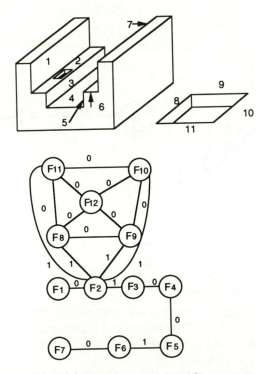

Figure 8.6 An example of feature interaction with common faces (Joshi and Chang, 1988).

1. Features interact in such a way that they only have common edges between them.
2. Features intersect in such a way that they share a common face, and interaction between the features splits a face of the feature.

An example of interaction with common edges between features is shown in Figure 8.6. This interaction may cause faces of one feature to be split up. During the recognition process, the split faces have to be merged into a single node. This procedure requires that:

1. The two faces have the same equation.
2. The two faces have one face to which both are adjacent.

When these two conditions are satisfied, the sub-graph corresponds to an interaction of features and not to a single feature. To recognize the constituent features, the sub-graph must be decomposed into primitive elements. A procedure called split-edges is used to separate the features. This procedure is:

1. Identify all the nodes with number of 1 incident arcs >1.
2. Delete all 1 arcs that emanate from such nodes.
3. Form components of the graph.
4. Add back 1 arcs deleted within each component.

But this heuristic procedure causes separation of features where no separation is necessary. In the example shown in Figure 8.6, the desired features are slot and pocket, but the heuristic yields the result step, slot, and pocket.

This problem is corrected by the procedure join-features. This procedure evaluates pairs of features to determine if they can be merged into a single feature. In this procedure, only the considerations such as adjacency and split face information, which are local, are used. The process merges features by merging the list of faces that make up the features, reconstructing the adjacency for the merged list. Then the recognizer determines if all the merged features can form a single feature.

In the common face type of interaction, the features split a face of the feature. An example of this kind of interaction is shown in Figure 8.7. In the part shown, the two slots share a common face. This type of feature can be separated by the procedure split-nodes. The procedure is shown in the following steps:

1. Find the nodes that correspond to the pair of split faces in the graph and assign to set A. This is done by checking the face equations from the B-Rep information. If the set A is empty, then the procedure is stopped.
2. Find the nodes adjacent to A with 0 arcs and assign to set B. If the set B is empty, then go to step 8.

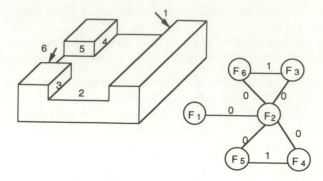

Figure 8.7 An example of features with common faces (Joshi and Chang, 1988).

3. Find the nodes that correspond to faces that cause the split and assign to set C. These nodes are adjacent to A with 1 arcs.
4. The node N to be split is adjacent to (AUC) with 0 arcs.
5. Create set D such that
   D ← C
   for each element in D.
   Find node n_j adjacent to it such that n_j does not belong to (AUN). Assign n_j to set D.
   Next element in D.
6. Connect the nodes in D to new node N' if the nodes are adjacent to N and delete 1 arcs that connect nodes in D to other nodes not in D.
7. Combine the split face node pairs in set A into a single node and upgrade the graph.
8. Re-initialize variables and repeat all the steps.

When this procedure is applied to Figure 8.7, the following results are obtained:

1. The faces 3 and 4 are identified as the split faces and assigned to set A.
2. The node 2 is adjacent to A with 0 arcs and hence is assigned to set B.
3. The nodes 5 and 6 are identified as the nodes that cause the split and are assigned to set C.
4. The node to be split is node 2.
5. The set D is created with the nodes 5 and 6.
6. The nodes in D are connected to 2' if they are adjacent to 2, and the arcs between 6 and 3 and between 5 and 4 are deleted.
7. The nodes 3 and 4 are merged into one node.

The components of the split graph are recognized as two slots.

REFERENCES

Carver, G. and Bloom, H.M. (1991) Concurrent engineering through product data standards. Manuscript for a chapter in *Control and Dynamic Systems: Advances in Theory and Application*, **35**, Academic Press, San Diego, CA.

Chang, T.C. (1990) *Expert Process Planning for Manufacturing*, Addison-Wesley Publishing Company Inc.

Clark, S.N. and Libes, D. (1992) FED-X The NIST Express Translator. National PDES Testbed Report Series NISIIR 4822, April.

Cunningham, J.J. and Dixson, J.R. (1988) Designing with features: the origin of features. *Computers in Engineering*, ASME.

Graves, G. *et al.* (1989) An interface architecture for CAD/CAPP integration using knowledge based systems and feature recognition algorithms. *International Journal of Computer Integrated Manufacturing*, **1**(2), pp. 89–100.

Groover M.P. and Zimmers, W. (1984) *CAD/CAM Computer Aided Design and Manufacturing*, Prentice-Hall Inc., Englewood Cliffs, NJ.

Houten, F.J.A.M., Erve, A.H. and Kals, H.J.J. (1989) PART – A Feature Based Computer Aided Process Planning System. Proceedings of the 21st CIRP International Seminar on Manufacturing Systems, Stockholm, Sweden, June 5–6, pp. 16–31.

Joshi, S. and Chang, T.C. (1988) Graph based heuristics for recognition of mechanical features from 3-D solid model. *Computer Aided Design*, **30**(2), March, pp. 58–66.

Joshi, S. and Chang, T.-C. (1990) Feature extraction and feature based design approaches in the development of design interfaces for process planning. *Journal of Intelligent Manufacturing*, **1**(1), pp. 1–15.

Kusiak, A. (1990) *Intelligent Manufacturing Systems*, Prentice-Hall International Series in Industrial and Systems Engineering.

Kusiak, A., Szczebicki E. and Vujosevic R. (1991) Intelligent design synthesis: an object oriented approach. *International Journal of Production Research*, **29**, pp. 1291–1308.

Lakko, T. and Mantyla, M. (1991) New Form Feature Recognition Algorithm. In *Computer Applications in Production and Engineering: Integration Aspects*, Elsevier Science Publishers, p. 369.

Martino, D.F. *et al.* (1991) *Computer Applications in Production and Engineering*, Elsevier Science Publishers, pp. 361–8.

Miller, R.K. and Walker, T.C. (1988) *Artificial Intelligent Applications in Engineering*, Prentice-Hall Inc.

Palmer, M. and Gilbert, M. (1991) Guidelines for the Development and Approval of Application Protocols, working draft version 0.7, ISO TC184/SC4/WG4, Document NI February.

Reed, K.A., Harrod, D. and Conroy, W. (1990) The Initial Graphics Exchange Specification, Version 5.0. National Institute of Standards and Technology, Interagency Report 4412, September.

Shah, J.J. (1992) *Features in Design and Manufacturing, Intelligent Design and Manufacturing*, John Wiley and Sons Inc.

Shah, J.J. and Rogers M.T. (1988) Functional requirements and conceptual design of the feature-based modelling system. *Computer-Aided Engineering Journal*, February.

Wilson, P.R. and Pratt, M.J. (1988) A Taxonomy of Features for Solid Modelling, Geometric Modelling for CAD Applications, IFIP.

CHAPTER 9

A survey of CAPP systems

9.1 INTRODUCTION

It has been almost three decades since computer techniques started to be used in assisting process planners' work. Various systems have been developed and many different implementation approaches have been employed for computer aided process planning systems. Many different surveys have been carried out by different institutions and researchers. This chapter is based on several previous well-known survey works and gives a general overview of computer aided process planning systems. Then, based on the overview, some widely known CAPP systems will be discussed in detail in the following section.

9.2 GENERAL SURVEY

Many surveys have been done in the last thirty years on the history of the development of CAPP systems. Many keynote papers have addressed many issues for the development of CAPP. Among these widely used keynote papers, we will point out some of them. Moseng addressed the geometrical model requirement for the future development of CAPP in his paper The Process Planner's Work Place Today and Tomorrow, which was published in *Robotics and Computer Integrated Manufacturing* (Moseng, 1984). Allen introduced the decision making scheme and process planning knowledge representation in the form of the constraint-based decision tree approach in his paper An Introduction to Computer-Aided Process Planning, which was published in the *CIM Review* (Allen, 1987). Tulkoff edited two general books with the titles *Computer Aided Process Planning* and *CAPP: From Design to Production* (Tulkoff, 1985, 1988): these two books collected some good research results from the CAPP field. Spur addressed the potential of using artificial intelligence techniques in his paper Advanced Methods for Generative Process Planning, which was presented in the first CIRP Working Seminar on Computer Aided Process Planning in Paris (Spur, 1985). Then, in 1987, the 19th CIRP International Seminar was hosted by Pennsylvania State University, where many researchers gave important reports based on their experiences. Merchant reported the role of CAPP within the CIM environment with the title CAPP in CIM – Integration and Future Trends (Merchant, 1987). Tulkoff reported a historical review and future prospects with the title Process Planning: An Historical Review and Future Prospects (Tulkoff, 1987). Koenig reported a bottom-up CAPP approach with the title Computer Aided Process Planning for Computer Integrated Manufacturing (Koenig, 1987). Tipnis addressed some development problems in his paper with the title Computer Aided Process Planning (Tipnis, 1987). Of course, these are only a few of the keynote papers which are considered to be guiding work in the development of CAPP.

In terms of survey work, the first comprehensive work was reported in 1982 in the 31st CIRP general annual meeting. The survey work was carried out by three internationally known scholars, Weill, Spur, and Eversheim, with the title Survey of Computer Aided Process Planning Systems, which was a keynote paper presented at the meeting. This survey work addressed almost all the technical problems for the development of CAPP, from process selection to the editing of process sheets. The survey work also reported the structures of systems for technology processing in the form of bar diagrams. Finally, about 16 different systems were reported (Weill *et al.*, 1982). This work was continued by Eversheim, and the second survey work was reported three years later at 34th CIRP general meeting with the title CIRP Technical Reports: Survey of Computer Aided Process Planning Systems. 56 systems were collected, and a general evaluation was given. This report concluded with Trends in Development, in which the authors addressed further developments, transferring variant systems to intelligent generative systems (Eversheim and Schulz, 1985). Within the same year, another comprehensive survey work was given, prior to the above work, in January 1985 at the 1st CIRP Working Seminar on Computer Aided Process Planning in Paris. This work was carried out by Wysk, Chang, and Ham with the title Automated Process Planning System – An Overview of Ten Years of Activities (Wysk *et al.*, 1985). This paper reported the survey for both variant and generative process planning systems. This survey work addressed the integration of process planning with the CAD model by means of a common part database on which current research is still focused. The report provided a comprehensive survey of about 29 systems in the form of a table. The table was further modified and published in Chang and Wysk's book *An Introduction to Automated Process Planning System* (Chang and Wysk, 1985). In the section of that book entitled A Survey of Process Planning Systems, seven well known systems of that time were introduced in detail. This book has been widely used recently for research work.

Another very widely used survey work was produced by Alting and Zhang with the title Computer Aided Process Planning: The State-of-the-art Survey, published in the *International Journal of Production*

Research in 1989 (Alting and Zhang, 1989). This paper is the most comprehensive one in terms of a survey of the CAPP field. More than 220 technical papers are referenced and 156 CAPP systems are reported in the appendix of the paper. This may be the only paper which represents so many developed systems. In this paper, the development of future trends was addressed in section 4, which also addressed some critical technical problems. Many of these issues have not yet been solved and are still focused on by current researchers. One year before, in 1988, Ham and Lu published their keynote paper for the 37th CIRP general annual meeting with the title Computer Aided Process Planning: The Present and The Future. This paper discussed theoretically the issues of planning, artificial intelligence in process planning, and the integration of process planning with CAD and production scheduling. This is one of the most comprehensive works in the development of process planning (Ham and Lu, 1988).

From the above discussion, we know that the past decade, the 1980s, was a very active period of development of CAPP systems and many survey works were written. The authors of this book have paid particular attention to collecting all reported CAPP systems, including traditional variant systems and up-to-date expert process planning systems. Figure 9.1, based on the authors' previous collection (Alting and Zhang, 1989), illustrates the system development's historical records up to 1989. Details of the reports in Figure 9.1 are represented in the

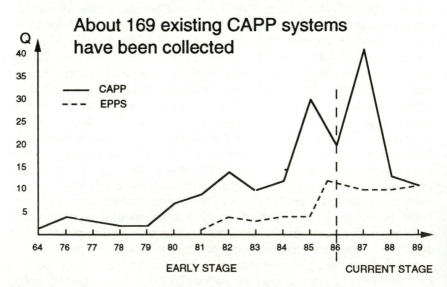

Figure 9.1 Historical record of development of CAPP and EPPS.

form of a table (Table 9.1). Table 9.1 might not include all of the CAPP systems that existed at that time, but it covers most: in total, 187 CAPP systems are reported. Most of the information for Table 9.1 was collected in 1989 by the authors. A few new systems were added recently, in 1992, including the first CAPP system developed by Novey in 1965 and the first CAPP system employing artificial intelligence required for the development of CAPP, called GARI, developed in France. The developers included were from all over the world, from North America to Asia. The programming languages covered range from traditional FORTRAN to up-to-date PROLOG and LISP, as well as C (the object-oriented language). The solid line in Figure 9.1 indicates the CAPP systems which may include both variant and generative expert process planning systems; the dashed line indicates the expert process planning system. It is worth mentioning that, in 1989, the authors collected only 14 new CAPP systems, and all 14 systems were AI based expert process planning systems. This fact indicates that the trends of using AI techniques had already been adopted by most CAPP developers. The other fact worth discussing is that new system reports declined after the peak of 1987. Figure 9.1 does not reflect new systems reported after 1989. The trends of new system reports do decline. Does this mean that trends of research and development of CAPP declined? No! New system reports declined because the development of CAPP has been sparked off all over the world, from developed countries to developing countries. Integrated production requires CAPP as a link to integrate design and manufacturing. Because of this, many systems have already been developed. Actually, many CAPP systems have already been used in real manufacturing environments. Some new systems may be developed, without being published in technical journals (most of the authors' information is collected from technical journals). On the other hand, most developers have moved their effort from developing new systems to maintaining the existing systems and exploring some difficult issues which have not been solved in the area of CAPP. These are the reasons why recent new system reports have declined.

In the following sections, a general survey is reported. This survey is based on more than 200 references, and 187 CAPP systems are listed in Table 9.1. Table 9.1 includes information such as spectrums of plannable workpieces, computer programming languages, input and output styles, interface possibilities, database methodologies, commercial availability, and developers of systems. For some systems complete information is not available (blanks indicate such incompleteness) and some information is listed in terms of percentage statistics.

Statistics of general features for these 187 existing CAPP systems are listed in Figure 9.2. Note that all of the expert process planning systems are also generative in nature.

Table 9.1 List of CAPP systems

Systems	Part shapes	Approaches	Characteristics & commercial situation	Programming languages	References & date	Developers
(1) ABTOTPZK	Rot.	Variant			Eversheim & Schulz (1985)	CIOMB (Bulgaria)
(2) ACAPS	Rot.	Semi-generative	Part family Coding	FORTRAN IV	1982 Emerson and Ham (1982)	Westinghouse Elec. Co. (USA)
(3) AC/PLAN	All	Variant and Generative	$		Apollo Computer catalogue (1986)	American Channels (USA)
(4) ACPSAP		Generative	KK-3 Coding Scheme		Wang and Wysk (1986)	Penn. State U. (USA)
(5) ACUDATA/ UNIVATION	Rot. and Pris.	Variant	Part number need		1978 Chang and Wysk (1985)	Allis Chalmers (USA)
(6) AGFPO	Forming process	Generative	Knowledge based	PROLOG	1985 Chryssolouris (1986)	Eshel et al. (USA)
(7) AIMSI	Rot.	Generative	Integrated with CAD/CAM		Wang and Wysk (1988)	State U. of New York (USA)
(8) AMPS	Sheet	Generative	Expert System Interfaced with CAD modeling	Common-LISP	1988 Inui et al. (1988)	U. of Tokyo (Japan)
(9) AMPS	Pris.	Generative	Expert System Developed for an integrated system QTC	C and LISP	1988 Chang et al. (1988b)	Purdue U. (USA)
(10) APLAN	Rot. and Pris.	Generative			Eversheim and Schulz (1985)	Gebr. Honsberg (W. Germany)
(11) APP	All	Generative	$ Interfaced with CAD, ES		Haas and Chang (1987)	GE and LA (USA)

	Type	Approach	Notes	Language	Reference	Institution
(12) APPAS	Milling and drilling	Generative	Surface Modeling Language (COFORM) used	FORTRAN IV	1977 Wysk (1977)	Purdue U. (USA)
(13) APS		Variant and Generative			Moseng (1984)	VTL/WZL (W. Germany)
(14) ARPL	All	Variant			Rembold and Dillmann (1986) p. 87	Tech. U. of Stuttgart (W. Germany)
(15) AUSPLAN	Rot.	Semi-generative	Based on the DCLASS System	FORTRAN 77	1987 Lin and Bedworth (1987)	Arizona State U. (USA)
(16) AUTAP	Rot. and Sheet	Generative	$ Interfaced with CAD		1975 Eversheim et al. (1980)	WZL (W. Germany)
(17) AUTAP-NC	Rot. and Sheet	Generative	$ Part program Interfaced with CAD		1978 Eversheim (1980, 1982)	WZL (W. Germany) RWTH
(18) AUTAP-Prism.	Pris.	Generative			Eversheim and Schulz (1985)	WZL (W. Germany)
(19) AUTOCAP	Rot.	Variant		BASIC	1977 Wright et al. (1987)	UMIST (UK)
(20) AUTODAK	Rot.	Generative	For single and multiple-spindle automatic lashes		1976 Weill et al. (1982)	IPK (W. Germany)
(21) AUTODYN Arbeitspl.	Rot.		$		Eversheim and Schulz (1985)	Hoff. and Partner (W. Germany)
(22) AUTOGAM					Weill (1982)	CETIM (France)
(23) AUTOPLAN	Rot.	Variant and Generative	$ Special language used for input	FORTRAN IV	1980 Wolfe (1985)	MRA (USA)
(24) AUTOPRO-GRAMMER	All	Variant	$		Eversheim and Schulz (1985)	Oerlikon-Boeh. (W. Germany)
(25) AUTOPROS		Variant	$		1966 Moseng (1984)	NAKK (Norway)

Table 9.1 Continued

Systems	Part shapes	Approaches	Characteristics & commercial situation	Programming languages	References & date	Developers
(26) AVOGEN	Rot.	Generative	$ ES	LISP	1989 Tonshoff et al. (1989)	U. of Hanover (W. Germany)
(27) BGCAP	All	Variant	Shaped surface input	BASIC	Jiang and Xu (1987)	BIME (China)
(28) BHCAP	Rot.	Variant	Shaped surface input	BASIC	Jiang and Xu (1987)	BIAA (China)
(29) BITCAPP	Rot.	Semi-generative	Primitive approach input	FORTRAN	Jiang and Xu (1987)	BIT (China)
(30) BPT Arbeitspl.	Sheet				Eversheim and Schulz (1985)	VEB Indus. KMS (W. Germany)
(31) CACAPSS	Rot.	Variant	GT code and shaped surface input	BASIC	Jiang and Xu (1987)	Sheryong No. 3 Machine Tool Works (China)
(32) CADCAM	Hole making	Generative	Extension of APPAS	FORTRAN	1980 Chang and Wysk (1985)	VIP and SU (USA)
(33) CAOS	Rot.	Generative	For single-spindle automatic lathe	FORTRAN	Huang et al. (1986)	North Western Polytechnical U. (China)
(34) CAP	Sheet	Variant	Part number needed		1963 Chang and Wysk (1985)	Lockheed-Georgia (USA)
(35) CAPE	All	Variant	Interfaced with CAD		Haas and Chang (1987)	Carrett Turbin (USA)
(36) CAPES	Pris.	Generative	Expert System CAD	C and Franz-LISP	1988 Fujita et al. (1988)	Mitsubishi El. Co. (Japan)
(37) CAPES	All	Variant	$		Eversheim and Schulz (1985)	Methods Workshop (UK)

(38) CAPEX	All	Variant			Eversheim and Schulz (1985) 1976	EXAPT (W. Germany)
(39) CAPP	All	Variant	$ User interactive	FORTRAN	Link (1976) Tulkoff (1978)	CAM-I (USA)
(40) CAPP-I	Rot.	Variant	GT code and shaped surface input	BASIC	Jiang and Xu (1987)	The 5th Ins. of Ministry of Machine Building (China)
(41) CAPSY	Rot.	Generative	Interactive input		1978 Spur et al. (1978)	IPK (W. Germany)
(42) CAR	Pris.	Variant			Akiba and Hitomi (1975)	Osaka Inst. of Infor. (Japan)
(43) CIMPP	Pris.	Generative	Knowledge based		Lai and Cai (1990)	Shanghai Jiao Tong U. (China)
(44) CIMS	Pris.	Generative	Knowledge based Interfaced with a CAD modeling named CIMS/MODE		1987 Iwata and Sugimura (1987)	Kobe U. (Japan)
(45) CIMS/PRO	Pris.	Generative	Special language CIMS/DES used for input		1980 Chang and Wysk (1985)	Kobe U. (Japan)
(46) CMPP	Rot.	Generative	$ English-like language (COPPL) used, interfaced with CAD/CAM	FORTRAN 77	1982 Sach Jr. (1982) Austin (1986)	UTRC (USA)
(47) COATS	Rot.	Generative	$ Tool selection for turning within the PICAP	PROLOG	Giusti (1986)	Pisa U. (Italy)

Table 9.1 Continued

Systems	Part shapes	Approaches	Characteristics & commercial situation	Programming languages	References & date	Developers
(48) COBAPP	Rot.	Generative	$ APPOIC code system used for input		1978 Chang and Wysk (1985)	Phillips/Purdue (USA)
(49) COMCAPP V	Rot. and Pris.	Variant	$ JCODE system used for input		Chang and Wysk (1985) Wolfe (1985)	MDSI (USA)
(50) COMP	Rot.	Generative	$ Expert System		Special report (1987)	Colding International Co. (USA)
(51) COPICS	All	Variant			Eversheim and Schulz (1985)	MTU (W. Germany)
(52) CORE-CAPP	Rot.	Semi-generative	GT codes used		Li et al. (1987)	Penn. State U. (USA)
(53) COROCUT	Rot.	Variant	$		Crocut Handbook (1986)	Sandvik (Sweden)
(54) C-PLAN	All	Variant	$		CADCentre Ltd (1987)	CADCentre Ltd. (UK)
(55) CPPP	Rot.	Generative	$ COPPL language used for interactive		1978 Nau and Chang (1983)	UTRC (USA)
(56) CRUNCH	Elec. Assemb.	Variant and Generative			Haas and Chang (1987)	Sperry Co. (USA)
(57) CSD and AML			$		Special report (1987)	Rath and Serang Co. Lexington (USA)
(58) CUTDATA	Rot. and Pris.	Variant	$		Special report (1987)	MRAT (USA)

System	Part type	Approach	Remarks	Language	Reference	Developer (Location)
(59) CUTPLAN	Rot. and Pris.	Variant	$ Only process determination		Tulkoff (1987)	MRAT (USA)
(60) CUTTECH	Rot.	Generative	$ Knowledge based For operation planning		1984 Barkocy and Zdeblick (1984)	MRAT (USA)
(61) CWOS-GPP	Rot. and Pris.	Generative			Haas and Chang (1987)	Texas Instruments (USA)
(62) DAPP	Rot. and Pris.	Variant and Generative	Interfaced with CAD		Haas and Chang (1987)	NBS (USA)
(63) DATASAAB	All	Variant			Eversheim and Schulz (1985)	Saab Scania, (Sweden)
(64) DCLASS	All	Variant and Generative	$ Tree structure system	FORTRAN 77	Allen (1980, 1987) DCLASS manual (1985)	B.Y.U. (USA)
(65) DIAL	Mech. Assembly	Generative			Baartman et al. (1990)	Delft U. (The Netherlands)
(66) DICAPP	Pris.	Generative	Knowledge based Interface with AutoCad	C	Sanii et al. (1989a)	North Carolina State Univ. (USA)
(67) DISAP	All	Generative	$ Extension of AUTAP		1981 Weill (1982)	WZL (W. Germany)
(68) DOPS	Drilling only	Generative	Interface with 3D-COMPAC	FORTRAN-LISP	Major (1987)	IPK (W. Germany)
(69) DPP	Pris.	Generative	Knowledge based Developed from KAPPS Integrated with a scheduling system	Common-LISP and C Smalltalk-80 SIMAN	Iwata et al. (1989, 1990)	Osaka U. (Japan)
(70) DREKAL	Rot.	Generative			1981 Weill (1982)	IFSW. U. of Hanover (W. Germany)

Table 9.1 *Continued*

Systems	Part shapes	Approaches	Characteristics & commercial situation	Programming languages	References & date	Developers
(71) EASE		Variant	$		Special report (1987)	Ease Co. Calif. (USA)
(72) ECAPP	Elec. Assembly	Generative	Knowledge based Developed from PWA-Planner	LISP	Fraser et al. (1990)	Digital Equipment Corp. (USA)
(73) EMAPS	All	Generative	$ Expert System		Tipnis (1986)	Tipnis Co. (USA)
(74) EPPSEA	Elec. Assembly	Generative	Expert System	TURBO PROLOG	Sanii et al. (1989b)	North Carolina State Univ. (USA)
(75) ESOP	Pris.	Generative	Expert System	FORTRAN and PROLOG	Tilley et al. (1989)	CRIT/WICM (Belgium)
(76) EXAPT	Rot. and Pris.	Variant	$ NC program		1964 Budde (1973)	EXAPT-Verein (USA)
(77) EXCAP	Rot.	Generative	Expert System Derived from AUTOCAP	PROLOG	1981 Wright et al. (1987) Davies (1984)	UMIST (UK)
(78) EXCAPP	Rot.	Generative	Expert System	PROLOG	1988 Du and Liu (1988)	BIAA (China)
(79) EXPLAN	Pris.	Generative	Expert System Interface with CAD	C and Kee shell	Warneche et al. (1989)	IPA Stuttgart (Germany)
(80) FAUN		Variant	NC Programming complex surface		Eversheim and Schulz (1985)	TU Budapest (Hungary)
(81) FLEXPLAN	Rot. and Pris.	Generative	Knowledge based Integrated with Scheduling		Tonshoff et al. (1989, 1990)	U. of Hanover (Germany)
(82) FFS		Variant	NC Programming sculptured surface		Eversheim and Schulz (1985)	Computer and Auto. Inst. (Hungary)

Name	Type	Approach	Description	Language	Reference	Institution
(83) FRAPP	Pris.	Generative	Knowledge based	Kee shell	Henderson and Chang (1988)	Arizona State U. (USA)
(84) GARI	Holes only	Generative	Expert System Rule-based	MACLISP	1981 Descotte and Latombe (1981)	Grenoble U. (France)
(85) GECAPP-PLUS	All	Variant	$		Wolfe (1985)	GE Co. (USA)
(86) GEMOS	All	Generative	Knowledge based		1986 Schaffer (1986)	MRAT and GE Co. (USA)
(87) GENPLAN	All	Variant and Generative	Based on CUTTECH $ Part family code used		Tulkoff (1987a,b)	Lockheed-Georgia (USA)
(88) GENTECH	Rot.	Variant			Eversheim and Schulz (1985)	CIOM (Bulgaria)
(89) GETURE	All	Variant	$		1975 Chang and Wysk (1985)	GE (USA)
(90) GIPPS	Pris.	Generative	Based on XIPLUS shell, interface with CAD, ES	C	Khorami (1988a,b)	U. of Strathclyde (UK)
(91) GLEDA-FT/-SZ	All	Variant			Eversheim and Schulz (1985)	Inst. of Tech. (Hungary)
(92) GLIM	Rot.	Variant			Holevi and Weill (1980) Weill (1982)	ITT (Israel)
(93) GPPS	Rot.	Generative	Expert system		1989 Peklenik and Sluga (1989)	U. of Ljubljana (Yugoslavia)
(94) GT-CAPP	All	Variant	Part family numbers used		Strohmeier (1987)	Rockwell Inc. (USA)
(95) GTIPROGE	Rot.	Variant			Eversheim and Schulz (1985)	Inst. of Tech. (Hungary)

Table 9.1 *Continued*

Systems	Part shapes	Approaches	Characteristics & commercial situation	Programming languages	References & date	Developers
(96) GTIPROGU and FM	Pris. and Sheet	Variant			Eversheim and Schulz (1985)	Inst. of Tech. (Hungary)
(97) HAL-T1 FT/-SZ	Rot.				Eversheim and Schulz (1985)	Hal-Rob. Ltd. (Israel)
(98) HICLASS	Pris.	Generative	$ Expert system Interfaced with CAD	C	1982 Liu (1985)	Hughes Aircraft Co. Calif. (USA)
(99) HIMAPP	Rot.	Generative	Artificial intelligence	Inter-LISP	Berenji (1986)	U. of S. Cal. (USA)
(100) HNDXCAPP	Pris.	Variant	Shaped surface input	BASIC	Jiang and Xu (1987)	Hunan U. (China)
(101) ICAPP	Pris.	Generative	Interfaced with CAD	FORTRAN	1981 Wright et al. (1987) Eskicioglu (1983)	UMIST (UK)
(102) IHOPE	Drilling Hole Only	Generative	Knowledge based		Voge and Zaring (1990)	Chalmers U. of Tech. (Sweden)
(103) INTELLI-CAPP	All	Generative	$ Combines expert system shell, GT, and related database		Tulkoff (1987a)	Cimtelligence Corporation (USA)
(104) IPROS	Rot.	Variant	COMPAC is used for input		1984 Moseng (1984) Rasch (1987)	APS (Ger.-Nor.)
(105) J2CAPP	All	Variant	GT code and primitive approach input	BASIC	Jiang and Xu (1987)	Jinan No. 2 Machine Tool Works (China)

Name	Type	Method	Notes	Language	Year / Reference	Location
(106) KAPLAN	Rot.	Generative	Knowledge based		1989 Giusti *et al.* (1989)	U. of Pisa (Italy)
(107) KAPPS	Rot. and Pris.	Generative	Know-how based Expert system	Common-LISP	1986 Iwata and Fukuda (1986, 1987)	Kobe U. (Japan)
(108) LOCAM		Generative	$ Expert system		Logan (1981) Tulkoff (1987a)	Logan Assoc. (USA)
(109) MASTER PARTS LIST	All	Variant			Haas and Chang (1987)	General Dynamics Co. (USA)
(110) MAYCAPP		Variant	$ Based on DCLASS	FORTRAN 77	1982 Zandin (1982) Wolfe (1985)	Maynard H.B. (USA)
(111) MIAPP	Rot. and Pris.	Variant			1979	OIR (USA)
(112) MICAPP	Rot.	Variant	$ Based on MICLASS	M-BASIC	Schaffer (1980)	Tennessee T.U. (USA)
(113) MICON	Rot. and Pris.	Variant			Sundaram and Cheng (1986)	T.U. Budapest (Hungary)
(114) MICRO-CAPP	Rot.	Variant	KK-3 coding scheme used		Eversheim and Schulz (1985)	Penn. State U. (USA)
(115) MICRO-GEPPS	Rot.	Semi-generative			Wang and Wysk (1986)	Penn. State U. (USA)
(116) MICRO-PLAN	Rot.	Generative	Knowledge-based Interfaced with CAD		Wang and Wysk (1986) Phillips (1987)	U. of Ill. at Chicago (USA)
(117) MIPLAN	Rot. and Pris.	Variant	$ Based on MICLASS		1976 Houtzeel (1976) Lesko (1983)	OIR and GE Co. (USA)
(118) MITURN	Rot.	Variant	$		1971 Koloc (1971) Bockholts (1973)	TNO and OIR (Netherl.-USA)

Table 9.1 *Continued*

Systems	Part shapes	Approaches	Characteristics & commercial situation	Programming languages	References & date	Developers
(119) MOPS	Pris.	Variant	Machining centre operation planning $	FORTRAN	Pinte (1987)	WTCM/CRIT (Belgium)
(120) MSA	Rot.	Variant	$		Haas and Chang (1987)	MSA (USA)
(121) MULTI-CAPP II	All	Variant	$ Based on MICLASS		1986 OIR Product News-Advance (1987) OIR Group Technology (1986)	OIR (USA)
(122) NITZLCAP	Gears	Variant	GT code and shaped surface input	BASIC	Jiang and Xu (1987)	NIT (China)
(123) OLPS		Variant			1975 Tulkoff (1981)	Boeing Co. (USA)
(124) OMS	All	Variant	Interfaced with CAD		Haas and Chang (1987)	Cummins Engine (USA)
(125) OPEX	Rot.	Generative	Expert System	PROLOG	1986 Gams et al. (1986)	FME Ljubljana (Yugoslavia)
(126) OPTAPLAN	Rot.	Generative	Knowledge based		Nordland (1987)	Mfg. Generative Systems Corp.
(127) OPGEN	P.c.b.	Generative	Expert system		1987	Hazeltine Corp. (USA)
(128) PART	Rot. and Pris.	Generative	Knowledge based selection based on XPLANE	FORTRAN 77	1989 Badiru (1988) Houten and Erve (1989)	U. of Twente (Netherlands)

System	Part type	Approach	Notes	Language	Reference	Institution
(129) PCCAPP	All	Variant	GT code input	BASIC	Jiang and Xu (1987)	JIT (China)
(130) PICAP	Rot.	Generative	Interfaced with COAST	PROLOG	1986 Santochi and Giusti (1986)	Pisa U. (Italy)
(131) PI-CAPP	All	Variant	$		1980 Chang and Wysk (1985)	Planning Insti. (USA)
(132) POPS	Sheet	Variant	Punching operation planning	FORTRAN 77	Pinte (1987)	WTCM/CRIF (Belgium)
(133) POPS	Mech. Assemb.	Variant			Haas and Chang (1987)	Hughes Aircraft (USA)
(134) POPULAR	Rot.	Variant and Generative	Interfaced with CAD		Sakamoto et al. (1987)	Komatsu Ltd. (Japan)
(135) PREPLA	Pris.	Generative			Rembold and Dillmann (1986) p. 92	Tech. U. of Stuttgart (W. Germany)
(136) PRICAPP	Pris.	Generative	IBM-PC XT		Pande et al. (1990)	IIT (India)
(137) PRIKAL	Pris.	Generative			Eversheim and Schulz (1985)	IFSW. U. of Hanover (W. Germany)
(138) PROPLAN	Rot. and Sheet				Marshall (1985a,b)	PERA (UK)
(139) PROTEMPS PS-System	All	Generative	$		Weill (1982) Eversheim and Schulz (1985)	ADEPA (France) PS-Technik (W. Germany)
(140)						
(141) PWA Planner	P.c.b. assembly	Generative	Expert System	PROLOG	Chang and Terwilliger (1988)	Purdue U. (USA)
(142) (P&WA)	Pris.	Variant	$		1977 Nilson (1977)	Pratt and Whitney Aircraft Co. (USA)

Table 9.1 *Continued*

Systems	Part shapes	Approaches	Characteristics & commercial situation	Programming languages	References & date	Developers
(143) QTC	Pris.	Generative	Integrated with CAD	C and LISP	Kanumury et al. (1988) Chang et al. (1988)	Purdue U. (USA)
(144) RATIBERT	All	Variant			Eversheim and Schulz (1985)	TH Otto V. Magdeburg (W. Germany)
(145) ROOPERT	Pris.	Generative	Knowledge based system	PROLOG and ALGOL 68c	1989	Cranfield Inst. of Tech. (UK)
(146) ROUND	Rot.	Generative		FORTRAN	Jared and Straed (1989)	Twente U. of Tech. (Netherlands)
(147) RPO/CAM	Rot.	Variant	$ Based on Autoplan	FORTRAN	Houten (1986) 1980 Vogel and Dawaso (1980)	MRAT and GE Co. (USA)
(148) RTCAPP	Pris.	Generative	Knowledge based Integrated with scheduling		Khoshnevis et al. 1989, 1990	U.S.C. (USA)
(149) SAGT		Variant			Chang and Wysk (1983)	Purdue U. (USA)
(150) SAPT	Rot. and Pris.	Generative	Based on hybrid concept of GT ES	PROLOG-86	1986	Beograd U. (Yugoslavia)
(151) SIB	Sheet	Generative			Milacic et al. (1987) 1982	Siemens AG. (W. Germany)
(152) SIPP		Generative	Expert System Semi-intelligent Frame-base	PROLOG	Weill et al. (1982) Nau and Chang (1985)	U. of Maryland (USA)
(153) SIPPS	All	Generative	Knowledge based	FORTRAN 77	Liu and Allen (1986)	Southampton U. (UK)

System	Type	Approach	Description	Language	Reference	Institution
(154) SIPS		Generative	Successor to SIPP Semi-intelligent	LISP	Nau and Gray (1986) Nau and Luce (1987)	U. of Maryland (USA)
(155) SISPA	Rot.	Generative	$ Based on SIB		Weill et al. (1982)	Siemens AG. (W. Germany)
(156) STAR	All	Variant			Haas and Chang (1987)	Grumman Co. (USA)
(157) SOPS	All	Variant	Sequence of operations planning	FORTRAN 77	Pinte (1987)	WTCM/CR17 (Belgium)
(158) System AV	All	Variant			Eversheim and Schulz (1985)	Microdata (W. Germany)
(159) System RW	All	Variant			Eversheim and Schulz (1985)	Weber Daten (W. Germany)
(160) TIPPS	Milling and Drilling	Generative	Knowledge-based Developed from APPAS & CADCAM interfaced with CAD		Chang and Wysk (1983, 1985)	VPI & SU (USA)
(161) TOJICAP	All	Variant	$ GT code and shaped surface input	BASIC	1982 Zhang et al. (1984)	Tongji U. (China)
(162) TOM	Rot.	Generative	Expert system Interfaced with COMPAC & EXAPT	PASCAL	1982 Matsushima et al. (1982)	U. of Tokyo (Japan)
(163) TOPS	Rot.	Variant	Based on DCLASS	FORTRAN 77	Pinte (1987)	WTCM/CRIF (Belgium)
(164) TRAUPROG-T	Rot.	Variant			Eversheim and Schulz (1985)	Inst. of Tech. (Hungary)

Table 9.1 Continued

Systems	Part shapes	Approaches	Characteristics & commercial situation	Programming languages	References & date	Developers
(165) TTR-S	Rot.	Variant		FORTRAN	1982 Hoffmann and Qarzo (1982)	Lang. Eng. Works (Hungary)
(166) TURBO-CAPP	Rot.	Generative	Knowledge-based Interfaced with CAD	PROLOG	1987 Wang and Wysk (1987a,b)	Penn. State U. (USA)
(167) TURN2	Rot. and Pris.	Variant	$		Special report (1987)	MICAPP Inc. (USA)
(168) VARGEN	All	Variant			Eversheim and Schulz (1985)	CRIF-K.U. (Belgium)
(169) VERDI	Rot. and Pris.	Generative	$		Eversheim and Schulz (1985)	IPA-TU Stuttgart (W. Germany)
(170) WICAPP	All	Variant	$ Based on DCLASS	FORTRAN 77	1982 Schwartz and Shreve (1982)	Westinghouse (USA)
(171) XCUT	Pris.	Generative	Knowledge based	Common-LISP	1989 Hummel et al. (1989)	Allied Signal Inc. (USA)
(172) XMAPP	Pris.	Generative	Expert System Interfaced with CAD Modeling	Common-LISP	1988 Inui et al. (1988)	U. of Tokyo (Japan)
(173) XPLAN	All	Generative	$ Expert system Based on DCLASS	FORTRAN 77	1984 Lenau and Alting (1986) Alting et al. (1988)	Tech. U. of DK. (Denmark)
(174) XPLANE	Pris.	Generative	Knowledge based	FORTRAN 77	1986 Erve et al. (1986)	Twente U. of Tech. (Netherlands)
(175) XPLAN-R	Rot.	Generative	$ Expert System Based on DCLASS	FORTRAN 77	1987 Zhang (1987; 1988a,b)	Tech. U. of DK. (Denmark)

No.	Name	Type	Approach	Description	Language	Reference	Institution
(176)	XPS-1	All	Variant and Generative	$ COPPL Used	FORTRAN 77	1982 Sack Jr. (1983) Groppetti and Semeraro (1986)	UTRC & CAM-I (USA)
(177)	XPS-E		Generative	Expert System Based on GRAI	MACLISP	Chryssolouris et al. (1986)	UTRC (USA)
(178)	ZCAPPS	All	Variant and Generative	$ Interfaced with CAD		Haas and Chang (1987)	Zeus Data Systems (USA)
(179)		Pris.	Generative	Interfaced with CAD modeling		1988 Kishinami et al. (1988)	Hokkaido U. (Japan)
(180)		Pris.	Generative	Knowledge based	C-PROLOG	Mill and Spraggett (1984, 1985)	Coventry Polytechnic (UK)
(181)		Pris.	Generative	Expert System Interfaced with CAD	Common-LISP	1988 Joshi et al. (1988)	Purdue U. (USA)
(182)		Rot.	Generative	Employed in FMS		1988 Kyttner et al. (1988)	Tallinn T.U. (USSR)
(183)		Mech. assembly	Generative	For roller chain assembly	PROLOG LISP/ Flavors	1989 Tonshoff et al. (1989)	U. of Hanover (W. Germany)
(184)		Mech. assembly	Generative	Knowledge based DEL Windows		Bartenschlager (1990)	ITR Germany
(185)		Sheet Metal	Generative	Knowledge based	TURBO-PROLOG	Uzsoy et al. (1991)	U. of Florida (USA)
(186)		Pris.	Variant	Knowledge based	PROLOG	Bandyopadhyay et al. (1987)	U. of Windsor (Canada)
(187)		Rot.	Generative	Knowledge based Clips-Shell ORACLE database Interface with scheduling	C	Zhang et al. (1992)	Texas Tech Univ. (USA)

General Features	Quantity	Percentage
Variant Approach	70	42.4 %
Semi-Generative Approach	5	3.0 %
Variant and Generative Approaches	10	6.1 %
Generative Approach	71	43.0 %
Expert or Knowledge Based Systems	39	23.6 %
Unknown	8	4.8 %
Interfaced with CAD Modelling	24	14.8 %
Using GT	16	9.7 %

Figure 9.2   Statistics of general features.

Plannable Workpieces	Quantity	Percentage
Only Rotational	53	31.2 %
Only Prismatic	16	9.7 %
Only sheet Metal	6	3.6 %
Only Holes	5	3.0 %
Both Rotational and Prismatic	16	9.7 %
Both Rotational and Sheet Metal	4	2.4 %
Both Prismatic and Sheet Metal	2	1.3 %
All (Rot., Pris., & Sh. M.)	40	24.2 %
Printed Wiring Board	1	0.6 %
Unknown	17	10.3 %
Mechanical Assembly	1	0.6 %
Electronic Assembly	1	0.6 %

Figure 9.3   Statistics of plannable workpieces.

Implementation Environment	Quantity	Percentage
BASIC	11	6.7 %
FORTRAN Family	27	16.4 %
LISP Family	11	6.7 %
PROLOG Family	12	7.3 %
C	5	3.0 %
PASCAL	1	0.6 %
Unknown	102	61.8 %
Shells and Toolkits	3	1.8 %

Figure 9.4 Statistics of implementation environments.

Plannable workpiece statistical results are presented in Figure 9.3. It is worth mentioning that CAPP systems for all plannable workpieces are usually framework systems, implying that further development work is usually necessary for application of these systems within real production environments.

Figure 9.4 shows the statistics of implementation environments. Some of the systems considered are bilingual; for example, some are implemented in both FORTRAN and LISP, some in C and Franz–LISP, etc. In this case, two corresponding items are taken into account.

Among these 187 CAPP systems, the systems which are commercially available cover about 31% and are marked with a $ symbol. It should be emphasized that this collection cannot be considered as encompassing all CAPP systems used throughout the world. Some systems may have escaped our attention when collecting data or were not reported for some reason. However, Table 9.1 is probably the first list covering so many existing CAPP systems. Figure 9.5 lists some of the famous process planning systems which use AI techniques. These systems are generally divided into two stages, the early and current stages. The early stage systems mainly emphasize how to implement process planning by means of artificial intelligence techniques. The current systems mainly emphasize increasing the integrability and intelligence of CAPP.

9.3 SYSTEMS SURVEY

In order to present the scope of existing CAPP systems and to demonstrate the characteristics of different systems which are implemented,

	NAME	DEVELOPERS	YEAR
E A R L Y S T A G E	CUTTECH	MRAT (USA)	1984
	EXCAP A	UMIST (UK)	1981
	GARI	Grenoble U. (France)	1981
	TIPPS	VPI & SU (USA)	1983
	TOM	U. of Tokyo (Japan)	1982
	XPS-I	CAM-I (USA)	1982
C U R R E N T S T A G E	EXCAP Y	UMIST (UK)	1989
	AMPS (1)	U. of Tokyo (Japan)	1988
	AMPS (2)	Purdue U. (USA)	1988
	DPP	Osaka U. (Japan)	1989
	PART	U. of Twente (Netherlands)	1989
	XCUT	Allied Signal Inc. (USA)	1989
	XPLAN-R	Tech. U. of Denmark (DK)	1988

Figure 9.5 Some examples of well-known expert process planning systems (EPPS).

18 well-known examples among these existing systems have been selected and have been discussed in alphabetical order. These 18 examples include variant systems, generative systems, and semi-generative systems, as well as expert systems. Some of them are illustrated in Figure 9.5.

9.3.1 APPAS, CADCAM and TIPPS

APPAS is an acronym for Automated Process Planning and Selection, developed by Wysk and presented in his dissertation at Purdue University (Wysk, 1977). It is probably the first well-known generative CAPP system written in standard FORTRAN, with a description of the detailed technological information of each machined surface represented by a special code. An implicit decision tree approach has been used to directly code the process capabilities in the system. APPAS is capable of selecting multiple passes and processes for the designed machined surface (such as twist drill–rough bore–finish bore); multiple diameter holes with special features such as an oil groove, slot or thread can be planned as a single machined surface. APPAS includes the selection of feed rate, cutting speed, diameter of the tool, number of milling cutter teeth, length of the tool, and length or depth of the cut for each tool pass. CADCAM is an extension of APPAS, developed by Chang and Wysk, which links APPAS with an interactive computer graphics terminal to demonstrate the concept of an integrated CAD and process planning system (Chang and Wysk, 1981). In a sense, TIPPS is a new generation of APPAS and CADCAM. TIPPS is an acronym for Totally Integrated Process Planning System, developed by Chang and Wysk at Virginia Polytechnic Institute and State University in 1982 (Chang and Wysk, 1983). It is probably the first system that integrates CAD and generative process planning into a unified system utilizing the AI and decision tree approaches. A special language called PKI (Process Knowledge Information) is used to describe the procedural knowledge (process capabilities). A CAD boundary representation is used as data input by the system. A user applies the crossbar cursor on a graphics terminal and a menu displays the surfaces to be machined in a process knowledge base to determine manufacturing processes, sequence, cutting parameters, and time estimation.

9.3.2 AUTAP and AUTAP-NC

AUTAP is an acronym for Automatisch Arbeits Planerstellung, developed at the Laboratory of Machine Tools and Production Engineering (WZL) at Aachen Technical University in West Germany (Eversheim *et al.*, 1980). It is one of the most complete generative process planning systems in use today. AUTAP-NC is similar to AUTAP: the only difference is that AUTAP is for the generation of process plans and AUTAP-NC is for the generation of part programs. Depending on the requirements of companies, the systems can be applied in an interactive or a batch mode. AUTAP and AUTAP-NC are parts of an integrated system for the generation of manufacturing documents which is already used successfully in different German companies. AUTAP works as follows: (1) determination of the raw material; (2) determination of the operation sequence; (3) selection of the machine tools for each operation; (4) calculation of the estimated time; (5) determination of the operation instruction. AUTAP-NC works as follows: (1) determination of manufacturing segments; (2) determination of tools and cutting data; (3) selection of lathe chucks; (4) determination of the manufacturing sequence; (5) generation of the part program. The system can handle different kinds of rotational parts like shafts, disks, rings, gear wheels, bearing caps, as well as sheet metal parts which are components for telecommunication equipment.

9.3.3 CAPP

CAPP is an acronym for CAM-I's Automated Process Planning system, developed by McDonnell Douglas Automation Company (McAuto) under a contract from CAM-I (Link, 1976). It is probably the first and also the most widely used of all process planning systems. CAPP is a database management system written in ANSI standard FORTRAN. It was developed primarily as a research tool to demonstrate the feasibility of computer assisted process planning, with logic based on Group Technology methods to classify and code parts. A structure is provided for a database, retrieval logic, and interactive editing capability. The coding scheme for part classification and output format are added by the user. A 36-digit (maximum) alphanumeric code is allowed. A coding scheme tailored to the user application is usually appropriate.

9.3.4 CMPP

CMPP is an acronym for Computer Managed Process Planning, developed by United Technologies Research Center (UTRC) in conjunction with the US Army Missile Command (Sack, 1982). It is one of the most technologically advanced systems in use today, written in FORTRAN 77 and currently running on both Univac and IBM systems. CMPP, which is the base for CAM-I's XPS-I and XPS-II, is a generative process planning system aimed at high technology machined cylindrical parts – parts characterized by expensive materials, tight tolerances, and complex manufacturing processes. CMPP addresses requirements such as: (a) turning, grinding, and honing to produce a cylindrical surface, (b) milling, electrical discharge machining and other non-cylindrical features, and (c) operations such as plating and heat treatment that apply to both cylindrical and non-cylindrical features. CMPP works in four steps: (1) generating a summary of operations; (2) selecting dimensioning reference surfaces for each cut in each operation; (3) determining and analyzing machining dimensions, tolerances, and stock removals for each surface cut in each operation; (4) generating a process plan output. An English-like problem-oriented language, COmputer Process PLanning (COPPL), is used to define manufacturing practices for families of parts. CMPP is interfaced to many CAD and CAM systems in American aircraft companies.

9.3.5 EXCAP and ICAPP

EXCAP is an acronym for EXpert Computer-Aided Process Planning, developed by Davies and Darbyshire at the University of Manchester Institute of Science and Technology (UMIST), UK (Davies and Darbyshire, 1984). It is an expert system for generating process plans for the machining of rotational components. Actually, EXCAP is developed from AUTOCAP, an earlier CAPP system from UMIST written in BASIC. Three generations of EXCAP prototypes have been developed; EXCAP-A, EXCAP-Y, and up-to-date EXCAP-P. The implementation language has been switched from BASIC to PASCAL and PROLOG, running on a SUN 3 under the UNIX operating system (Wright *et al.*, 1987). It is a rule-based system and uses the backward planning mechanism. It forms a tree of possible operation sequences. ICAPP is an acronym for Interactive Computer Aided Process Planning (Eskicioglu and Davies, 1983). It is a variant process planning system for prismatic parts written in FORTRAN on a VAX 11/750 under the VMS operating system. ICAPP is feature oriented and is capable of eight basic machining processes, namely face milling, peripheral milling, drilling, boring, reaming, tapping, counterboring, and countersinking. The architecture of ICAPP is illustrated in Figure 7.7. Recently ICAPP has been interfaced with CAD models via Initial Graphics Exchange Specification (IGES).

9.3.6 FLEXPLAN

FLEXPLAN was one of the earliest systems that attempted to integrate process planning with shop floor scheduling problems. FLEXPLAN was developed at the University of Hanover under the contract of the ESPRIT project 2457. The research work for FLEXPLAN concerned three main issues which were deeply embedded in the process planning area. They are (1) the development of a suitable representation of non-linear process plans, (2) the development of knowledge based methods for the automatic generation of non-linear process plans, and (3) the development of shop floor scheduling and rescheduling strategies for the utilization of non-linear process plans (Tonshoff *et al.*, 1989, 1990). Theoretically, FLEXPLAN can be supposed to work on both rotational and prismatic parts. FLEXPLAN works by attempting to model non-linear process plans in a computer and to generate these plans from a CAD database and thereby schedule and reschedule the shop floor in a flexible manner that permits frequent changes. FLEXPLAN generates a process plan that will have important alternatives for operation sequences and resources which are considered to support scheduling on the shop floor. It identifies manufacturing operations that can be carried out simultaneously, or independently of other operations. In order to represent a non-linear process plan, a three-layer concept for the representation of process plans is as follows:

Presentation layer

The presentation layer is the layer for the user interface. Process plans are displayed in a user-friendly way so that they can easily be read and modelled. This layer consists of a graph-based presentation with a condensed textual presentation.

Conceptual layer

The conceptual layer represents the process plan in much more detail. At this level, a process plan must represent a mathematical or abstract structure which can easily be processed by the system components.

Internal layer

The internal layer is the layer of translation. The conceptual representation of process plans is translated into a suitable data model so that the data can be communicated among different components of the FLEXPLAN.

This system provides the desired flexibility for planning a flexible manufacturing system environment, and the architecture of FLEXPLAN is illustrated in Figure 7.6.

9.3.7   GARI

GARI, probably the first AI based CAPP system to be reported in the literature, was developed by Descotte and Latombe at the University of Grenoble in France (Descotte and Latombe, 1981). GARI consists of a knowledge base of about 50 rules and a general purpose problem solver. Knowledge is represented by production rules dealing with conditions for the part being manufactured and advice representing technological and economical preferences. These preferences are weighted according to their importance and, where pieces of advice conflict, the system refers to their weights. The manufacturing rules are of the IF–THEN type and are parametrized by simple variables and set variables. GARI is implemented in the MACLISP language and operates on the CII-Honeywell BULL HB-68 computer under the MULTICS operating system. Recently GARI has been tested in the metal cutting industry with satisfactory results.

9.3.8   GENPLAN

GENPLAN is an acronym for GENerative process PLANning system, developed at Lockheed, Georgia (Tulkoff, 1981). It is one of the more sophisticated generative systems. It uses a Group Technology-based coding scheme that covers both part geometry and process variables to generate comprehensive operations. GENPLAN is oriented towards a specific environment. The initial GENPLAN was developed in 1976 based on CAM-I's CAPP system. At present, it has been split into two parts, Fabrication GENPLAN and Assembly GENPLAN. GENPLAN is being integrated with other companies' systems to increase the efficiency of the enterprise, such as interfacing it with a management decision support system, an automatic shop loading and scheduling system, and a tool order processing and accountability system.

9.3.9 KAPPS, CIMS, and DPP

KAPPS is an acronym for Know-how and knowledge Assisted Production Planning System, developed by Iwata and Fukuda at Kobe University in Japan (Iwata and Fukuda, 1987). It is probably the first CAPP system emphasizing special production know-how and experience. KAPPS basically consists of four sub-systems: (1) the CAD interface and user input sub-system; (2) the decision making sub-system; (3) the know-how and database; (4) the know-how acquisition sub-system. Know-how is represented and stored by using the proposed frame-base methodology. Know-how described by the production rules is also stored in the frame. The system can recognize the machined surface and the rough shape of the part, select the reference surfaces, determine precedence relations, select the machine tools, determine cutting conditions, and select the required cutting tools and fixtures. KAPPS was mostly written in the COMMON-LISP language and partially in SMALLTALK-80, and runs on a TEKTRONIX 440S system which has 5 Mbyte RAM and a 40 Mbyte disk unit. The architecture of KAPPS is illustrated in Figure 7.9.

From the reports, there is no apparent relationship between KAPPS and CIMS (Iwata and Sugimura, 1987). The reason why we put these two systems together is because they were developed under the same supervisor, Professor Iwata, when he worked at Kobe University. The two systems were developed in almost the same time period and use almost the same methodology. CIMS was reported a couple of years later than KAPPS. CIMS is a prototype system of the integrated CAD/CAPP systems. CIMS consists of an interactive modeling sub-system, a generative type process planning sub-system, and a knowledge base for process planning. The interactive modeling sub-system constructs the three-dimensional solid models of the parts, which include both geometrical information and technological information, such as the material data, surface roughness data, and accuracy data. The models are transferred to the process planning sub-system directly. Suitable sequences of machine tools and the machine sequence are generated based on the models of the finished part and the blank part constructed by the modeling sub-system. It is likely that while KAPPS emphasizes how to represent the know-how of process planning by means of knowledge representation in the knowledge base, CIMS emphasizes the integration of a CAD model which possesses some necessary manufacturing information concerning the process planning function.

The further development of KAPPS is concentrated on the integration of process planning with shop-floor scheduling, here in terms of the dynamic process planning (DPP) system. The proposed idea of DPP is required for determining both the process planning and scheduling simultaneously with the prepared alternative process plans. Input data of the DPP system are the loading and utilization status from the shop floor, the due date, production volume of each order from the material requirement planning system, and the product shape and production data from the CAD system (Iwata and Fukuda, 1989). Continuing, Iwata reported a new system named IPPS (Integrated Process Planning System) in 1990 (Iwata *et al.*, 1990). IPPS is probably the same system as KAPPS and has been further developed under the new name. The architecture of IPPS is shown in Figure 7.8. The idea of the DPP system is one step closer to real integrated manufacturing than the alternatives (or, in other words, non-linear process planning). Although the real DPP system has not yet been developed, the proposed prototype structure by Iwata is very promising.

9.3.10   MIPLAN and MULTI-II systems

MIPLAN is a variant process planning system developed in conjunction with the Organization for Industrial Research (OIR) (Houtzeel, 1976). The system was originally developed and implemented at the Lamp Equipment Operation of General Electric Company to obtain standard times for creating process plans. MIPLAN uses the MICLASS coding system for part descriptions. They are data retrieval systems which retrieve process plans based on part code, part number, family matrix, and code range. Recently OIR has announced their MULTI-II Group Technology system. MULTI-II is a comprehensive system that includes task-oriented modules: (a) MultiClass II Group Technology classification and retrieval; (b) MultiGroup II Group Technology analysis; (c) Multitreeve II design retrieval; (d) MultiCapp II computer assisted process planning; (e) MultiCats II automated time standards; (f) MultiTrack II tool tracking and inventory control. All of these modules utilize a single database (OIR Product News, 1987). MultiCapp II is a new version of MIPLAN.

9.3.11   PART family (ROUND, CUBIC and XPLANE)

PART is an acronym for Planning of Activities, Resources, and Technology and was developed at the Laboratory of Production Engineering at the University of Twente. The reason we give the title of the PART family in this section is that the PART system is integrated from four individual computer aided process planning systems, namely ROUND, CUBIC, XPLANE, and FIXES (Houten *et al.*, 1989, 1990; Erve and Kals, 1986). ROUND and CUBIC are twin sister systems for rotational and prismatic (cubic) parts respectively, and are based on almost the same concepts. ROUND and CUBIC models are executable programs which pass their results via a structured background memory file. When a module fails to deliver an acceptable result, this module or the proceding ones can be run again in order to obtain alternative solutions. XPLANE is the successor of ROUND, and the concept of the rule based expert system is introduced into the system. XPLANE has a compiled rule base which does not allow the process planner to change the rules during process planning: the change is restricted to the knowledge manager. FIXES is an auxiliary system which is focused on a systematic approach to the setup and work holding problems. FIXES is a fixture design expert system which is capable of selecting the most appropriate setups and their sequences of machining from an evaluation of the tolerances and orientations of the manufacturing features of a product. Based on the above four individual computer aided process planning systems, the PART project was initiated in 1988 in the form of a joint research project in which the Laboratory of Production Engineering and the Knowledge-based System Group from the computer science department of the University of Twente cooperated.

PART uses the GPM solid modeler for internal product representation. Models can be manipulated and edited via the volume editor. PART can be interfaced to every B-Rep modeler which supports the items mentioned above. It even has the capability to interactively add and edit tolerances and surface requirements after the recognition of the manufacturing features has been completed. In the PART project, Pro/Engineer is successfully used as an external feature based solid modeler. This has proved the value of the concept of decoupling feature based design and feature based manufacturing.

PART is implemented as a set of independently executing programs, called phases, which can be executed according to a user-defined scenario. The scenario describes the sequence in which the process and operations planning steps are executed by the supervisor. The phases can be distributed over a network of workstations and servers. This has two major advantages. First, it creates the possibility of achieving horizontal integration of CAPP functions in companies with large process planning departments; it is possible to allocate specific phases to specialists. Secondly, it makes it possible to borrow processing power from idling workstations. Workload balancing is done by the supervisor and can be influenced via the scenario. The execution of the phases can be monitored in a separate graphic supervisor window. The phases communicate with each other via the relational database in which all information on machine tools, tools, fixtures, features, scenarios, methods, strategies, etc. is stored.

9.3.12   PRICAPP

Among all the systems we have discussed in this section, PRICAPP, developed in the Department of Mechanical Engineering, India Institute of Technology, may be one of the youngest CAPP systems (Pande and Walvekar, 1989).

PRICAPP is a system that deals exclusively with the machining of prismatic parts used in the production of portable electric tools. A special shell was designed to realize the objectives of the system. This shell acts as an interpreter between the user and the operating system, which is DOS 3.2. A user-friendly interface package is developed on an IBM-PC/XT computer. The major part of the system is the planning module. The planning module consists of the component representation, machine selection, operation extraction and sequencing, process parameter selection, time calculation, and report generation. Typical data files used by the PRICAPP system are: the machine capability database that stores data about the machine features, such as speed, feed, tolerance, stroke, and dimensions; the tooling database that holds data about the tool code number, cutting edge length, maximum speed, and feed rate; and the machinability database that records the recommended speed and feed values for a particular tool or workpiece material. The non-cutting time library stores the non-machining times, such as speed change, coolant feed, and operator fatigue, taking into account the time/motion study constraints. The PRICAPP system has been found to be user-friendly and fairly inexpensive to implement. The architecture of PRICAPP is illustrated in Figure 7.10.

9.3.13   QTC

QTC is an acronym for Quick Turnaround Cell, developed by T. C. Chang at Purdue University. QTC is one of the earliest integrated process planning systems which is well integrated with CAD modeling (Kanumury *et al.*, 1988; Chang *et al.*, 1988; Kanumury and Chang, 1991).

QTC is an integrated system that enables rapid product prototyping. This system tries to avoid the intervention of the planner and even does the visual inspection after the part has been machined. The machining done is for prismatic parts. QTC consists of three primary modules, namely the CAD module, the Cell Controller module, and the Process Planning module. The CAD module accepts the part geometry from the user and then constructs the path using the Destructive Solid Geometry method. The Cell Controller module mainly schedules the part according to the Cutter Location data provided by the Process Planning module.

The interaction of the several different modules makes this system special in comparison with the others. The issues that are faced by this system are: interfacing with the design process, interfacing with the fixtures decision, interfacing with the cutter, interfacing with the shop floor databases, and providing effective decision support for the operation planning process. The system involves the close interaction of the user with the software and the hardware in the system. The system is UNIX based in a LAN protocol with the X-Windows system. The language is COMMON-LISP and C. The KEE expert shell has been used due to its powerful knowledge base capabilities. The system needs a large memory and is run on two Sun workstations. The architecture of QTC is illustrated Figure 7.11. QTC has been well explained by Chang in his book *Expert Process Planning for Manufacturing* (Chang, 1990).

9.3.14 SAPT

SAPT is an acronym for System for Automatic Programming Technology, developed by Milacic and Urosevic at the University of Beograd (Milacic *et al.*, 1986, 1987, 1988). SAPT is probably the first EPPS which involves Group Technology (GT) in the knowledge based system. The hybrid concepts of Group Technology (GT) and Type Technology (TT) are the basic philosophies of the system. The conceptual design is given by a processor with three segments: pattern recognition, functional logic, and optimization with economy. Based on the two main workpiece representation groups, rotational and prismatic, in GT terms SAPT is actually two systems: SAPT-R (for rotational parts) and SAPT-P (for prismatic parts). Both systems were developed on the same theoretical concept, with the same global structure. However, the same shell was used in two basically different manufacturing environments. The existing knowledge is built in as portions of different groups of manufacturing process entities. Access to the knowledge base may be achieved in two ways, by designing the manufacturing process either for a known workpiece or for a known machine tool. The first approach uses TT elements, and the other GT elements. The programming is accomplished in LISP and PASCAL, and analyses are made for applying the PROLOG language. The architecture of SAPT is illustrated in Figure 7.12.

9.3.15 SIPP and SIPS

SIPP is an acronym for Semi-Intelligent Process Planner, developed by Nau and Chang at the University of Maryland (Nau and Chang, 1985). It is a knowledge based system for the generative process planning of machined parts. It uses the frame-based knowledge representation system. The knowledge base consists of machinable surfaces, capabilities of various machining operations, a control structure which manipulates the knowledge, based on best-first branch and bound strategy, and it produces least-cost process plans, based on cost criteria that the user sets. The system is programmed in PROLOG and is currently composed of 55 frames. SIPS, a successor to SIPP, is an acronym for Semi-Intelligent Process Selector, written in the LISP language. Like SIPP, SIPS also employs the branch-and-bound least-cost-first search strategy. The basic difference between SIPP and SIPS is that SIPS uses a new knowledge representation technique called hierarchical knowledge clustering, instead of flat frames, to represent problem-solving knowledge (Nau and Gray, 1986). In SIPS, knowledge is divided into two categories: static knowledge and problem-solving knowledge. Static knowledge, which is internally stored as frames, is a representation of 3D objects. Problem-solving knowledge (knowledge about operation selection) is represented as hierarchical knowledge clusters, in which archetype and item frames are formed. Archetype frames are used to represent classes of machining operations. SIPS is currently being integrated into the AMRF (Automated Manufacturing Research Facility) project at the American National Bureau of Standards, where it will be used to produce process plans for an automated machine shop; and plans are under way for integrating it with systems being built at the General Motors Research Laboratories.

9.3.16   TOM, AMPS, and XMAPP

TOM is an acronym for Technostructure Of Machining, developed at the University of Tokyo in Japan (Matsushima *et al.*, 1982). It is a rule-based CAPP system written in PASCAL and runs on a VAX-11 computer. TOM was designed to accept input in two ways: (a) directly entering part description by the user, and (b) translating design data from the COMPAC CAD system. A front end for translating design data from COMPAC to the IGES data format was developed as an interface between CAD and TOM. The results are put into an EXAPT part program which is ready for execution. This allows for the integration of COMPAC and EXAPT. Because of the complexity of performing such translation, TOM handles 'holes' exclusively. TOM utilizes a backward-chaining mechanism in the process reasoning procedure. It begins with the goal state (the final state of a finished piece-part) and searches backward through the rule base to find operations required until the initial state (of the stock) is reached. Quite often, more than one rule qualifies for the selection. The method used in TOM to resolve the rule conflict is called the 'alphabet procedure'. The system is able to generate the optimal machining sequence based on the embedded knowledge.

AMPS and XMAPP were developed much later than the TOM system, but they were all developed under the supervision of Professor Sata at the University of Tokyo. AMPS is one of the few process planning systems for sheet metal bending. AMPS is an automatic bending process planning system which was developed based on the concept of integration with product modules and bending simulation techniques. This system requires the product model of the sheet metal part, which is represented as a set of plates and bends. The bending operation simulator is incorporated into the process planning system, which uses a homogeneous transformation to simulate the shape modification, springback, and manufacturing errors, according to the execution of bending. The simulation result is effectively used in the process plan generation. Expert system techniques are also introduced to utilize experts' knowledge in computers. The system represents such knowledge as production rules, and uses them as a heuristic to obtain the process planning results efficiently (Inui *et al.*, 1987a).

XMAPP was developed during the same time period in 1987 by almost the same developers at the University of Tokyo. XMAPP is particularly used for prismatic parts. The most important aspect of the XMAPP system is providing a product model. XMAPP is an automatic process planning system, which is developed based on the concept of the integration model of the machine part, especially with the form feature representation. XMAPP has the capability of automatic blank material designing, machining reference, clamping and positioning face determination, and machining operation planning. XMAPP represents process planning methods and constraints as production rules and applies them to product models. These results not only use product models for referring to the product information, but also dynamically modifies them as needed in the process of problem solving. XMAPP was applied in generating process plans for some machine parts to show its capabilities (Inui *et al.*, 1987b).

9.3.17 TURBO-CAPP

TURBO-CAPP, one of the most complex intelligent generative process planning systems to date, was developed by Wang and Wysk at Pennsylvania State University (Wang and Wysk, 1987a,b). The system consists of five modules: (1) machine surface identification; (2) process selection and sequence; (3) NC code generation; (4) knowledge acquisition, and (5) database management. Each module is composed of several routines undertaking a specific task. It works in the following steps: (a) interprets geometrical data from a 2½D CAD system (AUTOCAD), (b) extracts surface features in the CAD database, (c) checks design consistency of geometric dimensioning and tolerancing, (d) manages an extensive manufacturing knowledge base, (e) updates the knowledge base by interacting with experienced planners (learning), and (f) performs intelligent reasoning based on extracted machined surfaces and their relationships with surface finish, geometrical dimensions and tolerances, current machine configurations, and available tools, in order to generate alternative process plans. A combination of both frame and rule-based methodology is used to build up the knowledge base of the system. In brief, problem solving (process planning) knowledge is structured in three different layers: layer of facts, layer of inference rules, and layer of meta-knowledge for ease of reasoning. The system is implemented on IBM PCs in PROLOG.

9.3.18   XPS-I and XPS series systems and XCUT

XPS-I is an acronym for eXperimental Planning System, developed by the United Technologies Research Center (UTRC) under the Computer Aided Manufacturing-International Inc. (CAM-I) contract, and is the property of the CAM-I Process Planning Program members (Sack, 1983). It is the prototype of an advanced generative process planning system written in FORTRAN 77. XPS-I has the following major characteristics:

1. The system is manufacturer-independent – the user defines both the data requirements and the manufacturing logic for his own environment.
2. The system is generative.
3. The system is interactive – it provides user interactions for menu selection, data entry/display, and data file maintenances.
4. The system is customized.

XPS-I consists of three major area of methodology: (1) data dictionary; (2) decision modeling; (3) relational database. XPS-I is designed to provide a basic framework for advanced generative process planning. It is the first step in the development of a total system for the automated process planning of general parts. CAM-I's long range plans include the phased implementation of many additional capabilities, ultimately resulting in a complete system (XPS-n).

CAM-I's XPS series systems could not, in fact, be continuously developed according to CAM-I's original development schedule. In some literature XPS-II was mentioned (Chryssolouris, 1987), then XPS seemed to disappear from development. After a period of silence, a new system named XCUT appeared. It was said that XCUT was the succeeding development of XPS. This information was based on Ham's presentation at CIRP's workshop on CAPP in Hanover in 1987. From the literature report, XCUT was somewhat different from the XPS system. XCUT was developed by Hummel in Kansas City in 1988 (Hummel, 1988).

XCUT uses production rules, data tables, and a relational database. The output is in the form of an operation plan. The XCUT system architecture consists of four components: the user interface, the relational database, the expert planning system, and the solid modeler. The expert planning module is the essence of XCUT and has an inference engine, a knowledge base, and a working memory called HERB. There are approximately 500 production rules in XCUT that have to be considered. The system is implemented on LUCID Common LISP. The solid modeling system provides the part geometry information, the knowledge base; the system used is called ROMULUS. The user interface is achieved using the LUCID LISP Window Tool Kit. The workstation that is used to implement the system is the Apollo Engineering workstation. Much success has been reported using XCUT, yet there are many areas that may require some improvements and research is under way to perfect the system (Hummel *et al.*, 1989). The architecture of XCUT is illustrated in Figure 7.13.

9.4 EVALUATION OF CAPP SYSTEMS

In general, it is very difficult to evaluate a CAPP system absolutely. To evaluate CAPP systems requires knowledge of manufacturing process planning, computer software engineering, artificial intelligence, and so on. In a different time period, the evaluation of CAPP may emphasize different evaluation criteria. Recalling the early stages of the development of CAPP systems, most evaluations were focused on whether the systems could generate an executable process plan, and how much human interference was required, as well as how much CPU processing time was needed. Due to the rapid development of CAPP systems, the aspects of CAPP systems have changed dramatically compared with the early stage systems. The evaluation criteria have also changed. It is worth mentioning that the criteria of evaluation which we use today may not be exactly appropriate for tomorrow's CAPP systems. In this section we will try to provide some evaluation criteria that we often use today.

9.4.1 Criteria for evaluation

To evaluate a CAPP system, three aspects need to be considered: (1) the manufacturing process planning aspect; (2) the software aspect; and (3) the hardware aspect. Of course, some of the criteria may be categorized into all three aspects. The following criteria may be useful for evaluating current CAPP systems:

Capacity of planning workpiece

The capacity criterion includes the geometrical shape of the workpiece. For instance, some systems work for either rotational or prismatic or sheet metal, while some other systems may work on both rotational and prismatic workpieces. In addition, the dimensional size of a workpiece may vary from very small to very large, based on what kind of product the workpiece will eventually produce. The size of gears used in a wrist watch will be extremely different from the gears used in a 500 000 ton oil tank. While most current systems have not specified the size range of the workpiece, most of them focus on the middle range. Actually the process planning may be completely different when the sizes of the workpieces are different. This criterion has not been a concern in previous research.

The capacity criterion also concerns the variation range of the workpiece's geometrical shape. Workpieces in the real world my vary widely. For the category of rotational workpieces, some workpieces are very simple, while some others are extremely complex. The capacity of planning should be concerned about the ability of the system to deal with the complexity of workpieces. According to the survey of the work, most of the existing process planning systems can only deal with extremely simple workpieces, while some others claim that they can carry out relatively complex process planning.

Capability of planning workpiece

The capability of planning mainly concerns the details of the process planning carried out by the system. In Chapter 2, we discussed the process planning and operation sequencing. So far some CAPP systems can only generate process plans which do not include the detailed operation sequencing. This plan may be used for production scheduling but not for use in NC programming. The process plans that can be directly used for NC tapes must include the specific operation sequencing step by step. Some good systems even possess the function of operational tolerance calculation, which is very important in the allocation of cutting steps. The second concern of capability is that the process plan should not only include the detailed operation sequencing, but should also include the necessary tools, fixtures, and so forth.

Selectivity (intelligence)

This issue concerns the intelligence of the system. Apparently, the expert knowledge based or artificial intelligence based approach must be implemented into the system. Here the selectivity is mainly concerned with the process selection, tool selection, and fixture selection.

A good process planning system should possess the typical expertise from experienced process planners. This criterion should also include the calculation of time and cost for the process plans.

Consistency

One of the obvious merits of CAPP systems over human process planners is the consistency of the process plans. Even a very experienced process planner may carry out different process plans for the same workpiece at different times, and of course different planners will most often carry out different plans for the same workpiece. If the production rules in the knowledge base are not appropriate, and if the inference engine in the system is not suitable, the process plans carried out by the system may also be inconsistent. This will cause problems in production preparation.

Flexibility

Flexibility is the newest issue in evaluating a CAPP system. This criterion may seem to contradict the above consistency, but actually flexibility is one of the most important concerns in realizing the idea of CIM. This criterion requests the system to make process planning flexible based on the manufacturing resource availability. As we mentioned in the preceding section, some systems can generate alternative process plans based on the shop-floor equipment. This plan is called non-linear process planning or flexible process planning. Other systems attempt to carry out dynamic process planning, which is also referred to as just-in-time process planning. We discuss this issue in Chapter 10.

Integrability

Integrability is another criterion raised recently to measure the integrity of CAPP systems. With the rapid development of computer integrated engineering, the perception of final computer integrated manufacturing is closer than before. CAPP, playing a key role within the CIM environment, must be integrated with other manufacturing functions. The systems should be open and should easily communicate with other manufacturing systems. Data structure is required to be standard and as much manufacturing information as possible should be processed. Recent research is integrating process planning with CAD systems mainly by using feature based design and by integrating process planning with production planning and manufacturing resource planning (MRP II). The latter activity is identical to flexibility.

Portability

Most CAPP systems are developed based on a hardware platform. Some
are oriented to PCs, while some others are oriented to workstations. A
few systems are even implemented on mainframe machines. How to
make the systems removable and exchangeable is one of the criteria for
using them in a real manufacturing environment. Most manufacturing
companies have already used computers for different purposes. When
a CAPP system is going to be introduced into the manufacturing
environment, the system should be portable to the existing computers.

User friendliness

This issue mainly concerns the user-friendly interface, which will make
the system easy to use.

   Based on the above criteria, there are still some other criteria for the
evaluation of CAPP systems, which are database structure, operating
system environment, and cost.

9.4.2   Object-oriented evaluation

In different situations you may need to evaluate CAPP in different
ways. For different objects, you may use different criteria for evaluation. The idea of object-oriented evaluation needs to be discussed
here.

   For example, if you want to evaluate a CAPP system that is particularly for rotational workpieces, then the criterion concerning whether
the system can deal with prismatic or sheet metal workpieces is not
necessary in measuring the system. The only criterion you need to
satisfy is whether the system can sufficiently carry out process plans
that you are satisfied with. In other words, if you are evaluating a
process planning system for mass production, flexibility may not be as
important because once a machine is selected it will be engaged for a
long time. In addition, when you want to evaluate a CAPP system for
electronic assembly, then the selectivity may be completely different
from that which we discussed previously. The general idea of object-
oriented evaluation is that a CAPP system must be evaluated based on
the object to which the system is going to be oriented.

REFERENCES

Allen, D.K. (1987) An introduction to computer-aided process planning. *CIM
    Review*, Fall.
Alting, L. *et al.* (1988) XPLAN – An Expert Process Planning System and its

Further Development. 27th International Conference, UMIST, UK, 20–21 April.

Alting, L. and Zhang, H.C. (1989) Computer aided process planning: the state-of-the-art survey. *International Journal of Production Research*, **27**(4).

Austin, B.L. (1986) Computer aided process planning for machined cylindrical metal parts. *Autofact '86*, Detroit, Michigan, November 12–14.

Baartman, J.P. and Heemskerk, C.J.M. (1990) On Process Planning with Spatial Uncertainties in Assembly Environment. Proceedings of the 22nd CIRP International Seminar, Enschede, The Netherlands, June 11–12.

Badiru, A.B. (1988) Expert systems and industrial engineering. A practical guide to a successful partnership. *Computers in Industrial Engineering*, **14**(1), pp. 1–3.

Bandyopadhyay, S., Sengupta, A. and Dutla, S.P. (1987) The architecture of a knowledge based process planning system for small scale industries. *The International Journal of Advanced Manufacturing Technology*, **2**(3), pp. 37–48.

Barkocy, B.E. and Zdeblick, W.J. (1984) A Knowledge-Based System for Machining Operation Planning. AUTOFACT 6, Anaheim, California, October 1–4.

Bartenschlager, H. (1990) Automatic Process Planning for Assembly Lines using a Hybrid Knowledge Base. Proceedings of the 22nd CIRP International Seminar, Enschede, The Netherlands, June 11–12.

Berenji, H.R. (1986) Use of artificial intelligence in automated process planning. *Computers in Mechanical Engineering*, September.

Bockholts, P.A.J.M. *et al.* (1973) TNO MITURN, Programming System for Lathes. Proceedings of the Second IFIP/IFAC International Conference on Programming Languages for Machine Tools, PROLAMAT '73, Budapest, Hungary, April 10–13.

Budde, W. (1973) EXAPT in NC Operation Planning. 10th Annual Meeting and Technical Conference, New York City, USA, April 15–18.

Chang, T.C. (1990) *Expert Process Planning for Manufacturing*, Addison-Wesley Publishing Company, p. 281.

Chang, T.C., Anderson, D.C. and Mitchell, O.R. (1988) OTC – an Integrated Design/Manufacturing/Inspection System for Prismatic. Proceedings of the 1988 ASME Computers in Engineering Conference, San Francisco, CA, August 1–4.

Chang, T.-C. and Wysk, R.A. (1981) Interfacing CAD/automated process planning. *AIIE Transactions*, September.

Chang, T.-C. and Wysk, R.A. (1983) Integrating CAD and CAM through automated process planning. *International Journal of Production Research*, December.

Chang, T.-C. and Wysk, R.A. (1985) *An Introduction to Automated Process Planning Systems*, Prentice-Hall Inc., Englewood Cliffs, New Jersey.

Chryssolouris, G. and Groenig, I. (1987) Process Planning Interface for Intelligent Manufacturing Systems. Proceedings of 19th CIRP International Seminar on Manufacturing Systems, Pennsylvania State University, USA, June.

Chryssolouris, G. and Wright, K. (1986) Knowledge-based systems in manufacturing. *Annals of the CIRP*, **35**(2).

Davies, B.J. (1986a) Application of expert systems in process planning. *Annals of the CIRP*, **35**(2).

Davies, B.J. (1986b) Expert Systems in Process Planning. 7th International Conference on the Computer as a Design Tool, London, 2–5 September.

Davies, B.J. and Darbyshire, I.L. (1984) The use of expert systems in process-planning. *Annals of the CIRP*, **33**(1).

Davies, B.J., Darbyshire, I.L. and Wright, A.J. (1986) The Integration of Process Planning with CAD CAM Including the Use of Expert Systems. Proceedings of the International Conference on CAPE, Edinburgh, UK, April.

Descotte, Y. and Latombe, J.-C. (1981) GARI: A Problem Solver that Plans how to Machine Mechanical Parts. Proceedings of 7th International Joint Conference on Artificial Intelligence, Vancouver, Canada, August.

Du, P. and Liu, J. (1988) The Use of Expert Systems in Computer Aided Process Planning. Proceedings of the Seventh PROLAMAT Conference, Dresden, GDR, June 14–17.

Emerson, C. and Ham, I. (1982) An automated coding and process planning system using a DEC PDP-10. *Computers and Industrial Engineering*, **6**(2).

Erve, A.H. and Kals. H.J.J. (1986) XPLANE, a generative computer aided process planning system for part manufacturing. *Annals of the CIRP*, **35**(1).

Eskicioglu H. and Davies, B.J. (1983) An interactive process planning system for prismatic parts (ICAPP). *Annals of the CIRP*, **32**(1).

Eversheim, W. and Esch, H. (1983) Automated generation of process plans for prismatic parts. *Annals of the CIRP*, **32**(1).

Eversheim, W., Holz, B. and Zons, K-H. (1980) Application of Automatic Process Planning and NC-Programming. The CASA/SME Autofact West Conference, November.

Eversheim, W. and Holz, B. (1982) Computer aided programming of NC-machine tools by using the system AUTAP-NC. *Annals of the CIRP*, **31**(1).

Eversheim, W. and Schulz, J. (1985) CIRP Technical Reports: Survey of computer aided process planning systems. *Annals of the CIRP*, **34**(2).

Fraser, A., Sloate, H. and Tseng, M. (1990) ECAPP: Experiences with a Process Planning Tool using Artificial Intelligence. AUTOFACT Conference Proceedings, SME 1990, pp. 21–11 to 21–23.

Fujita, S. *et al*. (1988) Study of Practical Computer Aided Process Planning Based on Expert Systems. Proceedings of the Seventh PROLAMAT Conference, Dresden, GDR, June 14–17.

Gams, M. *et al*. (1986) OPEX – An Expert System for CAPP. 6th International Workshop on Expert Systems and their Applications, Avignon, France, April 28–30.

Groppetti, R. and Semeraro, Q. (1986) CAPP – Computer Aided Process Planning using Relational Databases. ISATA, Flims, Switzerland, October 6–10.

Haas, M. and Chang, T.-C. (1987) *A Survey on the Usage of Computer Aided Process Planning Systems in Industry*. Purdue University, USA, January 15.

Ham, I. and Lu, C.-Y. (1988) Computer-aided process planning: the present and the future. *Annals of the CIRP*, **37**(2).

Henderson, M.R. and Chang, G.J. (1988) FRAPP: Automated Feature Recognition and Process Planning from Solid Model Data. Proceedings of the 1988 ASME Computers in Engineering Conference, San Francisco, USA, August 1–4, pp. 529–36.

Hoffmann, P. and Garzo, A. (1982) The Process Planning System of Lang Engineering Works. Fifth International IFIP/IFAC Conference on Programming Research and Operations, May.

Houten, F.J.A.M., Erve, A.H., Jonkers, F.J.C.M. and Kals, H.J.K. (1989) PART, A CAPP System with a Flexible Architecture. Proceedings of CIRP International Workshop on Computer Aided Process Planning, Hanover University, Germany, September 21–22.

Houten, F.J.A.M., Erve, A.H., Nauta, J.M., Boogert, R.M. and Kals, H.J.K. (1990) PART, Selection of Machining Methods and Tools. Proceedings of the 22nd CIRP International Seminar on Manufacturing Systems, University of Twente, Enschede, Netherlands.

Houten, M. van (1986) Strategy in generative planning of turning processes. *Annals of the CIRP*, **35**(1).

Houtzeel, A. (1976) The MICLASS System. Proceedings of CAM-I's Executive Seminar – Coding, Classification, and Group Technology for Automated Planning, p-76-ppp-01, CAM-I Inc., Arlington, Texas.

Huang, N.K. *et al.* (1986) CAOS – A Generative Computer Aided Operation Scheming System for Single Spindle Automatic Lathe. Proceedings of the Winter Annual Meeting of the ASME, Anaheim, California, USA, December 7–12.

Hummel, K.E. (1988) Coupling Rule-based and Object-oriented Programming for the Classification of Machined Features. Proceedings of the 1988 ASME International Computers in Engineering, San Francisco, CA, August 1–4.

Hummel, K.E., Brooks, S.L. and Wolf, M.L. (1989) XCUT: An Expert Systems for Generative of Process Planning. Proceedings of International Industrial Engineering Conference, Toronto, Canada, May 12–18.

Inui, M. *et al.* (1987a) Automatic Process Planning for Sheet Metal Parts with Bending Simulation. Proceedings of ASME-WAM.

Inui, M. *et al.* (1987b) Extending Process Planning Capabilities with Dynamic Manipulation of Product Models. 19th CIRP International Seminar on Manufacturing Systems, Pennsylvania State University, June 1–2.

Iwata, K. and Fukuda, Y. (1986) Representation of know-how and its application of machining reference surface in computer aided process planning. *Annals of the CIRP*, **35**(1).

Iwata, K. and Fukuda, Y. (1987) KAPPS: Know-how and Knowledge Assisted Production Planning System in the Machining Shop. 19th CIRP International Seminar on Manufacturing Systems, Pennsylvania State University, USA, June 1–2.

Iwata, K. and Fukuda, Y. (1989) A New Proposal of Dynamic Process Planning in Machine Shop. Proceedings of CIRP International Workshop on Computer Aided Process Planning, Hanover University, Germany, September 21–22.

Iwata, K., Fukuda, Y. and Sugimura, N. (1990) A Proposal of Knowledge Based Structure for Integrated Process Planning System. Proceedings of the 22nd CIRP International Seminar on Manufacturing Systems, University of Twente, Enschede, Netherlands.

Iwata, K. and Sugimura, N. (1987) An integrated CAD/CAPP system with know-how on machining accuracies of process. *Journal of Engineering for Industry*, **109**, May.

Jared, G.E.M. and Stroud, I.A. (1989) A Knowledge Based Expert System with Geometric Reasoning Capabilities for Computer-Aided Process Planning. Proceedings of 21st CIRP International Seminar, Stockholm, Sweden, June 5–6, pp. 1:78–1:88.

Jiang, W. and Xu, H. (1987) CAPP Systems and Applications in China. 19th CIRP International Seminar on Manufacturing Systems, Pennsylvania State University, June 1–2.

Kanumury, M. and Chang, T.C. (1991) Process planning in an automated manufacturing environment. *Journal of Manufacturing Systems*, **10**(1), pp. 67–78.

Kanumury, M., Shah, J. and Chang, T.C. (1988) An Automatic Process Planning System for a Quick Turnaround Cell – An Integrated CAD and CAM System. Proceedings of USA–Japan Symposium of Flexible Automation, ASME, July 18.

Khoshnevis, B. (1990) Integrated Process Planning. Proceedings of Manufacturing International 90, ASME, Atlanta, Georgia, USA, pp. 243–8.

Khoshnevis, B. and Chen, Q. (1989) Integration of Process Planning and Scheduling Functions. Proceedings of 1989 IIE Integrated Systems Conference, Atlanta, Georgia, USA, pp. 415–20.

Khorami, M.T. *et al.* (1988a) Bridging the Gap Between CAD and CAM – GIPPS: A Generative Integrated Process Planning System.

Khorami, M.T. *et al.* (1988b) A Generative Integrated Process Planning System: Unexpected Material Benefits.

Kishinami, T. *et al.* (1987) An integrated approach to CAD/CAPP/CAM based on cell-constructed-geometric-model (CCM). *Robotics and Computer-Integrated Manufacturing*, **3**(2), pp. 215–20.

Koenig, D.T. (1987) Computer Aided Process Planning for Computer Integrated Manufacturing. Proceedings of CIRP International Workshop on Computer Aided Process Planning, Hanover University, Germany, September 21–22.

Koloc, J. (1971) MITURN: A Computer-Aided Production Planning System for Numerically Controlled Lathes. Proceedings of the Second International Conference on Product Development and Manufacturing Technology, University of Strathclyde, UK, April.

Kyttner, R. *et al.* (1988) Framework for Integrated Computer Aided Process Planning and Scheduling Systems. Proceedings of the Seventh PROLAMAT Conference, Dresden, GDR, June 14–17.

Lai, K. and Cai, J. (1990) CIMPP: A CIMS-Oriented CAPP System. Proceedings of the 22nd CIRP International Seminar, Enschede, The Netherlands, June 11–12.

Lenau, T. and Alting, L. (1986) XPLAN – An Expert Process Planning System. Second International Expert Systems Conference, London, UK, September 30–October 2.

Lesko, J.F. (1983) MIPLAN Implementation at Union Switch and Signal. The Association for Integrated Manufacturing Technology, 20th Annual Meeting and Technical Conference, April.

Li, J. *et al.* (1987) CORE-CAPP – A Company-Oriented Semi-Generative Computer Automated Process Planning System. 19th CIRP International Seminar on Manufacturing Systems, Pennsylvania State University, USA, June 1–2.

Lin, L. and Bedworth, D.D. (1988) A semi-generative approach to computer-aided process planning using group technology. *Computers Industry Engineering*, **14**(2), pp. 127–37.

Link, C.H. (1976) CAPP, CAM-I Automated Process Planning System. Proceedings of the 1976 NC Conference, CAM-I Inc., Arlington, Texas.

Liu, D. (1985) Intelligent Manufacturing Planning Systems. AUTOFACT '85, Detroit, USA, November 4–7.

Marshall, P. (1985a) The requirements of an integrated CAD/CAM system. *Computer-Aided Engineering Journal*, February.

Marshall, P. (1985b) Computer-aided process planning and estimating as part of an integrated CADCAM system. *Computer-Aided Engineering Journal*, October.

Matsushima, K., Okada, N. and Sata, T. (1982) The integration of CAD and CAM by application of artificial-intelligence techniques. *Annals of the CIRP*, **31**(1).

Merchant, E.M. (1987) CAPP in CIM – Integration and Future Trends. Proceedings of the 19th CIRP International Seminar on Manufacturing Systems, Pennsylvania State University, USA, June 1–2.

Milacic, V.R., Urosevic, M., Veljovic, A., Milev, A. and Race, I. (1987) SAPT –

Expert System Based on Hybrid Concept of Group Technology. 19th CIRP International Seminar on Manufacturing Systems, Pennsylvania State University, USA, June 1–2.

Milacic, V.R. and Urosevic, M.S. (1986) SAPT – Knowledge Based Expert System. International Conference on Computer-aided Production Engineering, Edinburgh, April.

Milacic, V.R. and Urosevic, M. (1988) SAPT – Knowledge based CAPP system. *Robotics and Computer-Integrated Manufacturing*, **4**(1/2), pp. 69–76.

Mill, F.G. and Spraggett, S. (1984) Artificial intelligence for production planning. *Computer-Aided Engineering Journal*, December.

Mill, F.G. and Spraggett, S. (1985) An artificial intelligence approach to process planning and scheduling for flexible manufacturing systems. *Computer-Aided Engineering Journal*, February.

Moseng, B. (1984) The process planner's work place today and tomorrow. *Robotics and Computer-Integrated Manufacturing*, **1**(3/4), pp. 237–44.

Nau, D.S. and Chang, T.-C. (1983) Prospects for process selection using artificial intelligence. *Computers in Industry*, (4), pp. 253–63.

Nau, D.S. and Chang, T.-C. (1985) A Knowledge Based Approach to Generative Process Planning. Symposium of Computer-Aided/Intelligent Process Planning, ASME, Winter Meeting, Miami Beach, Florida.

Nau, D.S. and Gray, M. (1986) SIPS: An Application of Hierarchical Knowledge Clustering to Process Planning. Proceedings of the Winter Annual Meeting of the American Society of Mechanical Engineers, Anaheim, California, USA, December 7–12.

Nau, D.S. and Luce, M. (1987) Knowledge Representation and Reasoning Techniques for Process Planning: Extending SIPS to do Tool Selection. 19th CIRP International Seminar on Manufacturing Systems, Pennsylvania State University, USA, June 1–2.

Nilson, E.N. (1977) Integrating CAD and CAM – Future Directions. The CASA/SME Advanced Computer Techniques for Design and Manufacturing Conference, April.

OIR Europe Inc. (1986) *Group Technology as CAD/CAM Integrator in the 80s*. OIR Europe Inc., brochures and information material, Lambertus Hortensiuslaan 74, 1412 GX Naarden, The Netherlands.

OIR Product News (1987) *Multi-II Group Technology System*.

Olson, W.W. and Devries, W.R. (1987) Logical Basis for Process Planning. 19th CIRP International Seminar on Manufacturing Systems, Pennsylvania State University, USA, June 1–2.

Park, M.W. and Davries, B.J. (1987) Integration of Process Planning into CAD using IGES. 19th CIRP International Seminar on Manufacturing Systems, Pennsylvania State University, USA, June 1–2.

Pande, S.S. and Walvekar, M.G. (1990) PRICAPP: A computer assisted process planning system for prismatic components. *International Journal of Production Research*, **28**(2), pp. 279–92.

Peklenik, J. and Sluga, A. (1989) Contribution to development of a generative CAPP system based on manufacturing process technology. *Annals of the CIRP*, **38**(1), pp. 407–12.

Phillips, R.H. *et al.* (1987) An Integrated Intelligent Design and Process Planning System. 19th CIRP International Seminar on Manufacturing Systems, Pennsylvania State University, USA, June 1–2.

Pinte, J. (1987) Computer Aided Process Planning. CAD/CAM 87, Teknologisk Institut, Denmark, October 20–22.

Sack, Jr. C.F. (1982) Computer Managed Process Planning – A Bridge Between CAD and CAM. The CASA/SME Autofact 4 Conference, November.

Sack, Jr. C.F. (1983) CAM-I's Experimental Planning System, XPS-I. Autofact 5 Conference Proceedings, Detroit, Michigan, November 14–17.

Sakamoto, C. *et al*. (1987) POPULAR – An Automatic Process Planning System. 19th CIRP International Seminar on Manufacturing Systems, Pennsylvania State University, USA, June 1–2.

Sanii, E.T. and Davis, R.E. (1989) Computer Aided Process Planning Using Object-Oriented Programming. Proceedings of 1989 IIE Integrated Systems Conference, Atlanta, Georgia, pp. 385–8.

Sanii, E.T., Lian, J. and Srinivasan, K. (1989) Computer Aided Process Planning for Electronics Manufacturing. Proceedings of 1989 IIE Integrated Systems Conference, Atlanta, Georgia, pp. 450–4.

Santochi, M. and Guisti, F. (1986) PICAP: A Fully Integrated Package for Process Planning of Rotational Parts. 18th CIRP Seminar on Manufacturing Systems, Stuttgart, FRG.

Schaffer, G.H. (1980) GT via automated process planning. *American Machinist*, May.

Schaffer, G.H. (1986) AI vs. conventional programming. *American Machinist and Automated Manufacturing*, August.

Spur, G. (1985) Advanced Methods for Generative Process Planning. 1st CIRP Working Seminar, Paris, January 22–23.

Spur, G. *et al*. (1978) CAPSY – A Dialogue for Computer Aided Manufacturing Process Planning. 19th International Machine Tool Design and Research Conference, UMIST, UK, September 13–15.

Strohmeier, A.H. (1987) Implementing computer-aided process planning: Rockwell International case study. *CIM Review*, Fall.

Sundaram, R.M. and Cheng, T.J. (1986) Microcomputer-based process planning using geometric programming. *International Journal of Production Research*, **24**(1), pp. 119–27.

Tilley, S., Pinte, J. and Peters, J. (1989) Expert Systems for Automatic Process and Operations Planning. Proceedings of CIRP International Workshop, Hanover University, Germany, September, pp. 141–68.

Tipnis, V.A. (1987) Computer Aided Process Planning: A Critique of Research and Implementation. Proceedings of CIRP International Workshop on Computer Aided Process Planning, Hanover University, Germany, September 21–22.

Tönshoff, H.K., Beckendrof, U. and Anders, N. (1989) FLEXPLAN – A Concept for Intelligent Process Planning and Scheduling. Proceedings of CIRP International Workshop on Computer Aided Process Planning, Hanover University, Germany, September 21–22.

Tönshoff, H.K., Beckendrof, U., Anders, N. and Detand, J. (1990) A Process Description Concept for Process Planning, Scheduling and Job Shop Control. Proceedings of 22nd CIRP International Seminar on Manufacturing Systems, Enschede, The Netherlands, June.

Tulkoff, J. (1981) Process planning in the computer age. *Machine and Tool Blue Book*, November.

Tulkoff, J., ed. (1985) *Computer Aided Process Planning*, Society of Manufacturing Engineers, Dearborn, Michigan.

Tulkoff, J. (1987) Process Planning: An Historical Review and Future Prospects. 19th CIRP International Seminar on Manufacturing Systems, Pennsylvania State University, USA, June 1–2.

Tulkoff, J., ed. (1988) *CAPP: From Design to Production*, Society of Manufacturing Engineers, Dearborn, Michigan.

Uzsoy, R., Ramcharan, D.J. and Martin-Vega, L.A. (1991) An experimental expert system for process planning of sheet metal parts. *Computers in*

Industrial Engineering, **20**(1), pp. 59–69.

Vogel, S.A. and Dawson, D. (1980) Integrated Process Planning at General Electric's Aircraft Engine Group. CASA/SME Autofact Conference, November.

Vogt, H.G. and Zaring, P.A. (1980) IHOPE: The Interactive Hole Operation Planning Expert. Proceedings of the 22nd CIRP International Seminar, Enschede, The Netherlands, June 11–12.

Wang, H.-P. and Wysk, R.A. (1986) Applications of microcomputers in automated process planning. *Journal of Manufacturing Systems*, **5**(2).

Wang, H.-P. and Wysk, R.A., (1987a) A Knowledge-based Computer Aided Process Planning System. 19th CIRP International Seminar on Manufacturing Systems, Pennsylvania State University, USA, June 1–2.

Wang, H.-P. and Wysk, R.A. (1987b) Intelligent reasoning for process planning. *Computers in Industry*, **8**.

Warnecke, G., Mertens, P. and Schulz, Ch. (1989) Artificial Intelligence in Computer Integrated Manufacturing. Proceedings of CIRP International Workshop, Hanover University, Germany, September, pp. 185–96.

Weill, R., Spur, G. and Eversheim, W. (1982) Survey of computer-aided process planning systems. *Annals of the CIRP*, **31**(2).

Wolfe, P.M. (1985) Computer-aided process planning is link between CAD and CAM. *Industrial Engineering*, **17**(8).

Wright, A.J., Darbyshire, I.L., Park, M.W. and Davies, B.J. (1987) EXCAP and ICAPP: Integrated Knowledge Based Systems for Process Planning Components. 19th CIRP International Seminar on Manufacturing Systems, Pennsylvania State University, USA, June 1–2.

Wysk, R.A. (1977) An Automated Process Planning and Selection Program: APPAS. Ph.D. Dissertation, Purdue University, West Lafayette, Indiana.

Wysk, R.A., Chang, T.C. and Ham, I. (1985) Automated Process Planning Systems: An Overview of Ten Years of Activities. 1st CIRP Working Seminar, Paris, January 22–23.

Zandin, K.B. (1982) Computer Aided Process Planning and Productivity. The CASA/SME 1982 International Tool and Manufacturing Engineering Conference, May.

Zhang, H.-C. (1987) *Overview of CAPP and Development of XPLAN*. Institute of Manufacturing Engineering, The Technical University of Denmark, Publication No. 87-26.

Zhang, H.-C. and Alting, L. (1988a) XPLAN-R: An Expert Process Planning System for Rotational Components. Proceedings of IIE 1988 Integrated Systems Conference, St Louis, Missouri, USA, October 30–November 2.

Zhang, H.-C. and Alting, L. (1988b) Introduction to an Intelligent Process Planning System for Rotational Parts. Proceedings of the Advances in Manufacturing Systems Engineering Symposium, ASME-WAM 1988, Chicago, Illinois, USA, November 28–December 2.

Zhang, H.-C., Huang, H. and Mei, J. (1993) An Integrated Model for Process Planning and Production Scheduling. Proceedings of the 1993 NSF Design and Manufacturing Systems Conference, Charlotte, North Carolina, USA.

Zhang, S. and Gao, W.D. (1984) TOJICAP – A system of computer aided process planning system for rotational parts. *Annals of the CIRP*, **33**(1).

CHAPTER 10

Crucial issues and future trends of CAPP development

10.1 INTRODUCTION

Due to the rapid development of computer techniques, the research aspects of development of computer aided process planning systems have been changed dramatically compared with the initial research activities. Many research issues have been carried out in the last three decades. AI techniques have been successfully introduced into computer aided process planning systems, so that the systems can have experienced process planners' expertise for making process plans. Manufacturing features can be recognized directly from a geometrical model, so that the CAD data can be transferred to a process planning model. Many generative process planning systems have been developed, so that automated process planning concepts can be realized. While all these promising results are being carried out, the original goal of the development of computer aided process planning still remains in the same direction – to integrate design and manufacturing. Yet some crucial issues have not been solved completely and have impeded the progress of development of automated process planning systems. In this chapter we are going to discuss some of these crucial issues for research interests and try to point out the near future trends of CAPP development. The first topic is about the challenges of CAPP research.

10.2 THE CHALLENGES OF CAPP RESEARCH

The issues involved in planning in the manufacturing environment require coordination between people and organizations, perhaps over time and across distance, and thus are often much more complicated than those involved in individual human planning. The difficulties are mainly due to the following factors (Ham and Lu, 1988):

- The designer's intention may not always be clear to the manufacturing engineer who will act on that intention. The respective languages used in their professions, the ways in which they express their intentions, their critical concerns, and their perspectives may all differ.

- Automation of process planning requires that part features can be automatically extracted from the product model without human interaction, but existing interfaces of CAD systems do not sufficiently consider this requirement of automated process planning. Needed information may be inaccessible or in an inappropriate form. Engineering drawings are the medium currently used by designers to communicate with manufacturing engineers, but these drawings may sometimes contain insufficient data, or the data may be hidden in forms which cannot be directly used, or extraneous data may be included which obscures the relevant information.

- The designer generating the drawings is often not aware of the constraints and limited resources that the manufacturing engineer has to deal with when carrying out these intentions. This may be due either to a lack of communication, or to the designer's lack of experience with production, or lack of information about the factory facility. Regardless of the reasons, it results in plans that cannot be smoothly executed, or which can only be executed with greater cost.

- The amount of time between the planning generation phase (at the design department) and the planning execution phase (on the shop floor) is normally much longer than that involved in individual daily planning. Due to the dynamic nature of a production environment, it is very likely that by the time a design is ready to be manufactured the constraints that were used in generating the plan have already changed greatly, and thus that plan has become less optimal or even totally invalid.

- The generation and execution of a complete production plan normally involve many different organization units, and may often span a long period of time and different geographic locations. These conditions make the plan-monitoring progress, critical for plan improvement, very difficult if not impossible. Without this feedback from the shop floor, it becomes difficult to measure the quality or goodness of a plan for future enhancement.

- In an automated environment, process and operations planning must result in data which have to deal with the utmost details and all these data have to be available before the actual manufacturing starts. Interpretation, adaptation, and completion of data in a later phase are not applicable.

The above difficulties are real challenges faced by researchers attempting to develop computer aided process planning systems. Some research efforts have already been devoted to addressing these difficulties. However, we feel that they will not result in significant impact unless means are devised to deal with these difficulties in a cohesive and integrated fashion. Furthermore, there are three interrelated aspects of the subject that must be addressed cooperatively by researchers working in the process planning area in order to achieve the goal of integrated planning:

Automating existing planning activities

Computer systems must be developed to assist and/or automate some portion of planning activities according to the way in which they are currently being performed by human beings. The majority of the present research in manufacturing planning falls into this category (Eversheim and Schulz, 1985).

Anticipating future planning challenges

The requirements for manufacturing planning in future factories must be anticipated, and planning techniques developed to meet these future needs. Although this has been recognized as an important activity, to date there are only a few research efforts aimed in this direction. The progressive introduction of new automated manufacturing systems, such as flexible manufacturing cells or systems, leads to task expansion in process planning. A functional correlation between the degree of automation and the necessary planning effort is obvious.

Suggesting a more logical planning structure

Suggestions should be made, based on the characteristics of various computer automation technologies and planning approaches, on the most logical ways of conducting process planning in practice, so that it becomes more automatable by computers. This requires a flexible use of different planning scenarios and the generation of alternative solutions on a cost-selective basis. These suggestions can be viewed as the feedback from computer based automation technology to process planning. This challenge, perhaps the most significant in terms of its cost-benefit return, is, unfortunately, the most neglected one. But it is obvious that process planning faces a number of influences which will change the contents of the planning procedure.

10.3 HARDWARE ASPECTS

Hardware development has come full circle from the earliest implementations of CAPP systems. As the digital computer first attained widespread acceptance, the only available machines possessing the requisite power for a task such as process planning were mainframe systems which featured individual workstations.

As computer technology progressed, the personal computer moved to the forefront and most CNC manufacturing shops featured machines on the shop floor controlled by independent personal computers that could be linked to a centralized database by means of a local area network (LAN). This provided adequate control for individual machines, but failed to accommodate the information that ultimately became available in developing CAPP software packages.

When engineering workstations first appeared in the early 1980s, many engineers and software researchers found them to have advantages for setting standards, performing multitasking by means of 32-bit architecture, running UNIX operating systems, and so on. Engineering workstations perform multitasking, the concurrent execution of several programs or program parts, making the typical workstation a suitable platform for development of automated process planning systems. In particular, many workstation vendors have supplied CAD/ CAM software already loaded onto the workstations, which provides considerable convenience for CAPP development. That is one of the reasons why engineering workstations are so popular for use in CAPP development.

However, selection of a platform for the development of automated process planning is still not entirely focused on current engineering workstations. Currently, CAPP software development is scattered over a wide spectrum, depending on the purpose for which it is primarily intended. The selection of platforms is also based on the environment where the CAPP systems are going to be employed. In most recent manufacturing environments minicomputers are widely employed for different purposes, from scientific calculation to production management. Many software packages are installed within the existing machines. If the development goal is going to integrate these existing software packages and to integrate different production functions, it may be wise to choose existing minimachines as your platform, or at least the development should be portable to minicomputers. This offers only a slight advantage in terms of machine control, but is truly the only way that complete CIM integration can be achieved. The same CPU and database may be accessed by all other company departments, such as personnel, finance, materials management, marketing, etc. A protocol for determining the hardware requirements for a facility interested in implementing a CAPP system is provided in Sherif and Heaney (1990). Some cellulate cells are integrated by means of workstations. If the goal of CAPP is to be integrated into the cell control, then obviously the workstation will be the best choice. As we have mentioned before, many CAD/CAM programs have already been incorporated into workstations. If your aim is to integrate your CAPP using these CAD/CAM programs, then no other choice is needed for the workstation. If however the purpose of the development is to solve specific research issues and the scope of the research is not too wide, then PCs may be a better choice in terms of economy. Recently, development of new features of PCs has dramatically increased their capacity and capability. The current Intel 80486 CPU, with 32-bit architecture and 25–66 MHz operating speed, has made the PC a very good choice for CAPP development. Now that new PCs with Intel's Pentium microprocessors operating at over 66 MHz and UNIX operating systems are available, PCs will probably become the first choice again for CAPP development.

10.4   SOFTWARE ASPECTS

Software progress has paralleled that of computer hardware and CAPP technology. FORTRAN was the language many early CAPP systems were coded in, due to its ability to handle calculations and familiarity in the engineering community. According to the survey we have discussed in Chapter 9, a number of systems are still FORTRAN based despite some advances made in artificial intelligence class languages such as LISP and PROLOG. With the great demand for intelligent process planning systems, LISP and PROLOG programming languages have received more and more attention. Many systems are implemented in LISP or PROLOG, but to reduce the implementation workload and time, the use of available expert shells and toolkits will be more convenient for developers. Some commercially available shells have already been listed in Chapter 7. However, the development of updated process planning systems does not only require an increase in the intelligence of the systems but also needs to gain much integrability, which will require object-oriented programming techniques. Object-oriented programming has been around for nearly 20 years, since the development of SMALLTALK (Goldberg and Robson, 1983), but it only really gained widespread interest as a scheme for developing programmed systems in the middle of the 1980s. Object-oriented programming combines data types needed to define a computational entity and the methods (executable code) necessary to manipulate the object values; in other words, each object contains its own data and the procedures that operate on that data. This is in contrast with conventional programming, which treats data and procedures as distinct and separate entities. Since an intelligent and integrated process planning system has to be integrated with some object-oriented database for necessary manufacturing information when the system makes decisions, object-oriented programming techniques are even more suitable for CAPP development.

Another important aspect of software for CAPP development is migratory ability. With the rapid development in computers, the boundary between mini/micro and mini/mainframe is diminishing. In addition, one clear and growing trend in the engineering market that is reflected throughout the computer industry is the movement toward distributed computing. The widespread integration of PCs, workstations, file servers, minicomputers, mainframes, and even supercomputers connected to a network, is helping the network business grow faster than all other computer segments by 40% to 45% each year (Nilssen, 1989). This indicates that many manufacturing enterprises and engineering institutions are implementing a comprehensive engineering network as a central part of their competitive computing strategy. In order to enable the development of process planning software which can communicate with other software packages throughout the network, software with a wide migratory ability and good communicatability must be kept in mind.

10.5 USER FRIENDLY INTERFACE

The user friendly interface is one of the criteria for evaluating a CAPP system. However, there is not a standard yet for measuring the level of a user friendly interface. In order to plan the user friendly interface, it is necessary to clarify which users will interface with CAPP the systems. In general, there are three different types of users who need to interface with CAPP systems:

Process planners who possess valuable domain expertise for process planning that needs to be transferred to the knowledge base of the system.

Knowledge engineers who are in charge of implementation of the knowledge base and the production rules and who are supposed to be very good at system development, knowledge acquisition, and representation.

System operators who are mainly working on system operation and who are supposed to know the operating system very well.

Actually, in the real development environment, the job tasks are not as clear as the above categories. For example, in some cases the process planner can also be a system operator, and a knowledge engineer may also be required to have some process planning expertise. Unfortunately, one cannot learn so much different knowledge with one limited brain. Even if one person could do three jobs as above, it would be very time consuming. In most cases it is just impossible. So the use of user friendly interfaces will reduce the burden of doing three different jobs and may eventually replace one or two jobs. The user friendly interface is mainly focused on two concerns: (1) process planning knowledge acquisition and modification, and (2) system access and communication.

Process planning knowledge acquisition and modification mainly involve knowledge engineers, as we have discussed above. The user friendly interface at this level is an internal layer interface, requiring communication between knowledge engineers and the knowledge base in an easy manner. The decision rules may be input and modified in terms of English-like or even English language. Ideally, during the use of the system, the knowledge engineer should be required as little as possible. At best, the knowledge engineer can eventually be replaced by the process planner, who may directly put his expertise into the knowledge base without requiring any programming experience. This interface level means that the knowledge base can be accessed by anyone who can modify the production rules, without any programming background.

System access and communication are mainly interfaced with system operators, as we have discussed above. The user friendly interface at this level is an external layer interface. It requires interactive communication between system operators and operating systems in an easy manner. At this level, interactive access to the system can be driven by means of a user menu, where the English-like or English language operation is also required. With the rapid development of operating systems and window facilities, the system operator can be anyone who can directly access and operate the system by means of a mouse. The operating commands may be entered by pressing the mouse buttons.

From the above discussions, we can see that the user friendly interface will dramatically reduce the number of users and simplify the process of operation. The person who accesses the system needs to know nothing about artificial intelligence, knowledge bases, programming languages, or operating systems and hardware.

10.6   INTEGRATION ABILITY

We discussed the user friendly interface in the preceding section. In this section we are going to discuss the integration ability for the development of CAPP systems. First, to avoid confusion between interface and integration, we would like to present the definition of the difference between interface and integration given by Ham and Lu (1988): One difference between interfacing and integration is that interfacing can be achieved at the result level while integration must be addressed at the task level. In other words, it would be too late to integrate a task when its sub-results (such as design and manufacturing specifications or decisions for process and operation planning) have already been decided separately. To achieve truly integrated design and manufacturing, the interactions between them should be addressed at a much earlier stage than that of our current focus. If these activities can be integrated at the task level, their differences will diminish gradually as we approach the result stage and their results will be naturally integrated. Similarly, various interrelated sub-tasks in manufacturing planning must be integrated at the task level, rather than be interfaced at the result level after they have been carried out separately.

The reasons for the integration of process planning are:

- Improved efficiency in the information flow
- Improved quality of the process planning
- Reduction of human errors
- Functional integration of process planning and scheduling, enabling a quick search for alternative solutions for optimization in the use of equipment and production control
- Flexible use of different functions

Also, the implementation of the integrated system must be based on:

- A uniform product description based on proper features
- The use of different modules for different functions
- The use of a uniform user interface for every module
- The use of a uniform database interface for every module
- The possibility of facilitating user interaction at the request of the operator

Process planning systems, in order to provide the integration, automation, and flexibility needed in future process planning, should:

- be generative
- be technology based
- use features as a technological and communicational interface between design and process planning
- be able to automatically extract all product data
- use a supervisory control system to ensure user friendliness and flexibility in use
- integrally support all planning tasks, including capacity planning and scheduling
- take decisions based on optimization techniques (also required for knowledge-based systems)
- be fit for closed-loop planning

A framework for integrated planning should have a multi-dimensional perspective. In other words, integration is needed in the following areas:

- **Planning knowledge:**
Science-based principles must be integrated with experience-based knowledge. Physics must be integrated with heuristics.
- **Planning activities:**
Process planning must be integrated downwards with operation planning and upwards with production planning. Operation planning must be integrated with physics (or models) of the manufacturing process being planned.
- **Planning techniques:**
Techniques such as group technology, modeling and simulation, optimization, and knowledge-based approaches must all be integrated into a truly robust planning system. The approach of building separate planning systems based on a single technique (e.g. rule-based systems) and then interfacing them with other systems which were built upon different techniques is not sufficient for integrated planning.
- **Planning constraints:**
Various planning constraints, local or global, technical or non-technical, user-provided or expert-provided, should be integrated during the planning stage, rather than being added to the system afterwards. This requires viewing manufacturing planning as a cooperative problem solving activity which incorporates various concerns simultaneously.
- **Planning feedback:**
Mechanisms must be provided to automatically incorporate feedback from the planning results to improve future planning decisions. The closed-loop planning is fundamentally different from the current open-loop planning architecture. It requires the extension of present planning activities to the plan monitoring and execution stages.

The following three sections describe three critical integration efforts: (1) integration of CAD with CAPP, (2) integration of CAPP with CAM, and (3) integration of CAPP with job shop-scheduling.

10.6.1   Integration with CAD

As we have discussed in Chapter 8, the concepts of feature based design provide a convenient approach for converting CAD data into process planning. The ability to analyze information visually is a fundamental task in integrating design and process planning. This task can generally be recognized as two major activities, namely model decomposition and feature recognition. Once the manufacturing features are recognized, the relevant manufacturing processes can be applied. There are two approaches to the use of features in process planning. One approach uses the global feature information for planning. In this approach, GT code is used to represent the feature on a manufacturing workpiece. The second approach takes individual features on the workpiece and determines the process and sequence for each of them. In this case, individual features and relationships between features must be identified and represented so that the system can automatically apply the decision rules according to the recognized features. This is exactly what current generative process planning systems perform. However, process planning is such a complex task that it requires not only features information but also other manufacturing information such as property of material, production size, surface roughness, hardness, as well as positioning tolerance and sometimes even the operation tolerance. Unfortunately, this information is inherent with features and cannot be represented in the form of current geometrical models. This is the problem of feature based techniques, which indicates that feature based techniques alone cannot be sufficient for providing all necessary manufacturing information for process planning and ensuing manufacturing functions. From this perspective, no common product model exists for the whole company. The proposed modeling of a product, including all necessary information for manufacturing, is a basic requirement for the integration of CAD and CAPP. This means that CAD must provide an element-oriented user interface which allows the input of technical elements such as holes, pockets, chambers, and grooves. Technical elements can be used by a designer in describing the part, or by a process planner in planning the operations to manufacture the part. The user interface should consist of input macros which define such technical elements as the relation of lines, face, or volume primitives. They should access standards as well as data for tools, fixtures, and machines.

For example, if the designer wants to place a groove in a shaft, he should be able to choose a specific macro for that groove. He should be able to specify the standard sheet to refer to and the face of the shaft where he wants to position the element. Depending on the diameter and length of the cylindrical face, all parameters of the groove should be determined automatically and shown on the screen by graphic and textual output. The designer should be able to change every single parameter of the groove.

These tasks of design and process planning indicate the direction in which computer aids must go in the future:

- open, extendable solid modeling systems
- a common product model for all applications
- a common factory database for all computer systems within a company

Nearly all company parameters can become important for the planning and execution of manufacturing processes. Every workpiece parameter should be available on the computer and, due to its object-orientation, should be part of the product model. This leads to the demand for arbitrary extendability of the model. Product models should be manipulatable with commercially available program modules as well as with user written programs. Algorithms for solid model manipulation must come as well-documented subroutine programs. Software vendors must not be reluctant about disclosing the initial data structure of their products.

10.6.2   Integration with CAM

The conditions for integrating process planning with computer aided manufacturing systems originally focused on converting process planning data into NC machine codes. At the very early stage, using computers to assist process planners' work was considered as an integral part of computer aided manufacturing (CAM), simply because the main job of process planning belongs in the manufacturing spectrum. That is why, at the beginning, we often heard the terms CAD/CAM but seldom heard of computer aided process planning. The term computer aided process planning (CAPP) generally appeared frequently when considerable efforts were put into that area, and lately it is considered a separate area from computer aided manufacturing. That is why we recently acquired the term CAD-CAPP-CAM. Integrating process planning with NC path has achieved great progress and is probably the first achievement for integrating CAPP with other manufacturing functions. In the early 1960s a numerical control programming language EXAPT, which is extended from APT, was developed and eventually linked with process planning functions in former West Germany. EXAPT was first interfaced with the GT criteria and was then interfaced with variant CAPP (Budde and Zolzer, 1973). Research attention in the area of development of CAPP systems was mostly focused on how to generate NC paths. Several good examples have been carried out recently, such as QTC, XMAPP, SPAT, and so on. Whether it is a variant or a generative CAPP system does not make much difference for integration with the NC path. The real work here is just how to convert the data which has already been produced from a process planning scheme into executable NC codes. In another words, integrating process planning with the NC path is just a different format of output from process planning. In general, a good CAPP system should be able to carry out both edited process planning by word processing software and an NC path which will be directly sent to NC machines. However, generating only an NC path will not lead to real automation because real production preparation not only requires an executable NC path but also needs information about tooling, fixturing, material handling and raw material stock. These are usually available in process planning, but are not necessary in an NC path. Scheduling information is also important for shop floor control, making the integration of CAPP become even more complex. In general, further research in the area of integration with CAM has extended into two categories. In order to convert all manufacturing information from CAPP to production control, the integration of CAPP with manufacturing resource planning (MRP I) has now become an activity. The research emphasis on this area is in trying to build a single database which can be shared by different manufacturing departments, such as the job floor, raw material storage, the material handling department, and the tooling and fixturing department, or even the personnel department. The other research emphasis is focused on integration of CAPP with job shop scheduling. This is mainly concerned with the critical information for job shop control, which we discuss in the following section.

10.6.3 Integration with production scheduling

Problems

Due to its complexity, process planning is often carried out without consideration of job shop status information, such as resource availability. Besides, throughput of orders in a job shop often suffers from disruptions caused by stochastic bottlenecks, non-availability of tools, or breakdown of equipment. Replanning has to be done by improvisation and can result in long through-put times. A large number of process plans perhaps cannot be executed and have to be altered. The process plan containing a linear sequence of operations is not flexible enough. A process plan that provides several possible ways to manufacture a given workpiece would be of great help. This shows the necessity of breaking away from the process plan as a static and linear operation sequence. Plan rigidity has to be overcome by enriching the plan representation.

Assuming that process plans for mechanical parts can be derived automatically from design databases, traditional process planning seems to be too highly abstract. Too many constraints referring to the workpiece and the job shop are ignored. A new process plan representation has to be developed which covers the causal structure of the manufacturing process. Instead of pruning all alternative sequences of manufacturing operations considered during planning except the 'optimal' one, it is proposed to raise the pruning process to a higher level. The plan representation should be able to express parallelism and alternative operations. It should support decision making on the shop floor by providing dependency information referring to the job shop status. Job shop control can benefit from such 'intelligent' plans, which can be adapted to the actual resource constellation. Non-linear, directed graphs are suitable to hold workpiece states as preconditions and specific manufacturing operations as events. The similarity to rule-based knowledge representation enables the methodical integration into artificial intelligence systems.

A process plan can, for instance, be represented by a Petri Net which provides the power to model logical and temporal relationships between manufacturing operations. Petri Nets easily represent concurrency and can contain loops; therefore, the modeling of interaction is possible. We can represent the actual state of the job shop as part of the planning knowledge representation by Petri Nets, but since the number of objects relevant to the job shop state is huge, a hierarchy of Petri Nets must be created which, taken together, will provide enough modeling power. Different objects (e.g. resources such as machines and tools) in the job shop can be modeled in separate Petri Nets. There must be some communication facility for the Petri Nets which make up the hierarchy. This can be achieved via token passing through global elements of the hierarchy.

Since it is possible to represent job shop status and process plans (which are part of the current set of shop orders) by Petri Nets, both representations can be merged for job shop control purposes. Assuming that a valid job shop schedule exists, control of operations in the job shop according to schedule is possible by utilizing the task-bidding principle. The basic idea is that, in order for a machining operation to take place, the required machine and tools must be available and in the proper status. This translates directly into the requirement that the corresponding Petri Nets modeling the resources must have a distinct token distribution. Simple task dispatchers based on this principle can be built to dispatch shop orders, tool preparation orders, transport orders, etc.

Job shop control algorithms and strategies should be developed to prove that shop control benefits from non-linear, directed graph process plans. The requirements of new algorithms for control purposes have to be defined. Such algorithms have to be developed and tested so that non-linear process plans can be utilized optimally. Strategies, or job scheduling and order dispatching, have to be evaluated.

The borderline between planning and control (i.e. scheduling) is fuzzy. Planning and control depend on each other and must ultimately use the same data. However, today's systems for production planning, scheduling, and job shop control do not take these dependencies into account. In a factory environment, a CAPP system, typically depending on heavy interaction with a process planner, is utilized for process plan generation. The order-independent process plan is stored in a database of active process plans. The process plan itself consists of a linear sequence of operations. Once an order is initialized by the PPS

(Production Planning and Scheduling System), the corresponding process plan is retrieved from the database, associated with the order data and, thus, a shop order is generated. At the scheduled release time, the shop order is stored in the database which contains all released shop orders. The job shop control system manages job shop operations. It accesses the shop orders database and updates the job shop status database. There is occasional feedback from the job shop control system to the PPS system. Systems are generally implemented in procedural programming languages. This architecture has some severe flaws, however:

- Job shop status is not considered during process plan generation. Thus plans are generated which cannot be processed because of the actual state of the job shop.
- When a process plan is retrieved from the database of active plans, it is often months, if not years, old and therefore possibly outdated.
- Since there is no support for replanning, the job shop control system cannot adequately react to disruptions. In the most primitive case, such a reaction could be the retrieval of a contingency process plan from the process plan database and its utilization instead of the original plan.
- Knowledge, which plays an important role in the realm of production planning, is very difficult to represent if procedural programming languages are employed.

Effectively, there is a complexity barrier which hinders the automation of more complex tasks like generation of operation sequences or schedules.

An advanced system which integrates process planning scheduling, and the job shop control system should manage all functions. Such a system should be knowledge based and possibly implemented utilizing AI software technology. This makes the integration of knowledge and, therefore, the higher automation of functions like process plan generation easier. An integrated system is monolithic only from a logical point of view; its implementation may very well have a distributed nature. It could work like this: once an order is received, the corresponding process plan is generated automatically, taking into account the actual job shop status. For this, we propose the term just-in-time process planning because the plan is generated on a when-needed basis. The resulting shop order is stored in the database of released shop orders. Job shop control is exerted by the integrated system already mentioned. Replanning in case of disruptions is facilitated because process plans can be altered with respect to the changed job shop status. Such an architecture can be realized if two preconditions are fulfilled:

- The integrated knowledge based system is powerful enough to allow automated generation of process plans
- Increased computing power is available

There will still exist a database containing all active process plans. This is for pragmatic reasons. Process plans are needed not only for scheduling and job shop control tasks, but also for a variety of other tasks. For example, in the area of investment planning, the task of machine procurement requires production times, set-up times, and descriptions of process per operation from a large number of process plans. Typically, these tasks involve many hundreds or thousands of active process plans from the database. Of course, if automatic process plan generation is available, as assumed here, it would also be possible to generate all necessary process plans on a when-needed basis, but this would consume excessive computing power. So, the strict correctness of all process plans in the database with respect to actual job shop status is sacrificed here for the sake of prudent use of computer resources.

Three different process planning approaches

On the basis of the above discussion, we are going to introduce three research approaches (Larsen and Alting, 1989):

1. non-linear process planning (NLPP),
2. closed loop process planning (CLPP), and
3. distributed process planning (DPP).

Non-linear process planning (NLPP) got its name because of its inherent property of being non-linear, i.e. the plans created do not form a linear structure; instead, they are branched at every node (Figure 10.1). NLPP can also be referred to as flexible process planning, in which the process planning mechanism takes a task and prepares all the possible plans for the task. This results in numerous optional

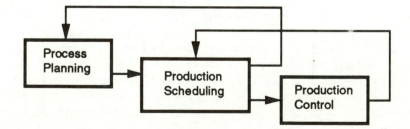

Figure 10.1   Configuration of non-linear process planning (NLPP).

plans. More often, plans themselves are ranked on the basis of their feasibility. Hence, it can be seen that before the part actually enters the manufacturing system all possible plans for its manufacture exist. The underlying assumption is that all problems which can be solved ahead of time, or which can be foreseen ahead of time, should be solved to reduce the simultaneous decision making requirements. When decisions are called for, all that is required is the selection of the plan rather than the creation of one. This helps in bringing some form of standardization. The disadvantage is felt when there are large numbers of parts in the system. Plans tend to increase exponentially with an increase in parts, which can cause a storage problem. Also, large numbers of unfeasible plans are created, according to the real time situation from the shop floor.

This approach is suitable for the conditions where an intelligent planning module is absent or where there is no method of automatically feeding the real time information back to the planning module.

Closed-loop process planning (CLPP), shown in Figure 10.2, forms a loop round which the information is passed until it is found consistent. CLPP can also be referred to as dynamic process planning. Once the consistency of the information is verified, plans are made according to the information in the loop. In the NLPP approach, due to the lack of availability of feedback facilities, plans are created and tried for suitability. In the CLPP approach, plans are created only upon feedback. The process planning mechanism will create process plans based on what the production plan tells it. Production planning often informs the process planning module about the availability of resources. Based on what resources are available, process plans are made. Therefore, every plan made is a feasible plan with respect to the production facilities available. This approach takes into consideration the dynamic behavior of the manufacturing system. It is essential to know the real time status. Production control informs production planning about the status of each resource, and the latter in turn informs process planning. Here, real time status is transferred all the way to process planning.

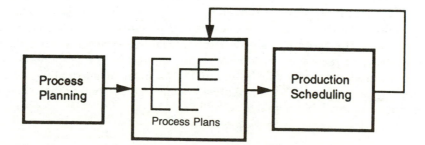

Figure 10.2 Configuration of closed-loop process planning (CLPP).

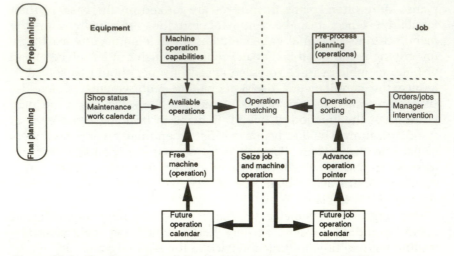

Figure 10.3   Configuration of distributed process planning (DPP).

Another very important observation is the response to real time situations. As the approach proposes feedback, it is possible to create a dynamic planning module by means of real time.

This approach makes it possible to create plans when they are needed and which are suitable to the present conditions. It is most appropriately used in a sequential request–answer form of manufacturing system, because all the process plans created are constrained by the shop status and ordered jobs.

Distributed process planning (DPP) is shown in Figure 10.3. The distinctive feature of this type of planning is its inherent integrated approach to the creation of a plan. This approach is commonly referred to as generative architecture. DPP carries out production planning and process planning simultaneously. The process plans are created based on the availability of resources. This approach divides the planning problem into two phases, pre-planning and final planning. Pre-planning is a phase where the job requirements are analyzed. This phase deals with identification of all resources required to machine a part, such as operations, fixtures, tooling, etc. Once all the requirements are known, DPP enters the second phase wherein available resources are matched with required resources. Flexibility of this approach can be increased if the required resources can be given in the form of alternatives. The more the alternatives for the required resources, the better the chance of matching available resource to it.

Integrated process planning model (IPPM)

Here, we are going to introduce an integrated process planning model (IPPM) which integrates process planning and production scheduling. The most important feature of this model is that it is a combination of closed-loop process planning and distributed process planning. From the previous discussion of process planning, it is clear that CLPP aims at real time response while DPP aims at resource matching. For achieving true integration, it is necessary to match resources and at the same time create plans in real time mode. Real time information only asks for a plan to be executed, but for the plan to be feasible it is necessary that an effective information system be available. This information is provided by a database. The architecture of the model is shown in Figure 10.4.

In this approach, CLPP updates all the individual functions about the real time situation on the shop floor. It also makes sure that all the information is collected at the required time and is available when needed. DPP is used for the purpose of generating the plan. This makes sure that the resource available matches the job requirements. Finally, the scheduling function assigns the job to the machines.

The machines are arranged in cells and these cells are capable of completing the jobs almost completely. Each cell is controlled by a cell controller, which is an executer as well as a monitor. It monitors information like: (1) machine status: which machine is in use at the particular time, did any machine break down? etc.; (2) tool status: which tools are being used on which machines, which tools broke? etc.; (3) labor status: what is the status of labor attendance? etc. All this information is stored in a database called the shop floor database which is updated regularly. An entry is made to it every time status changes on the shop floor. The function of production control is to retrieve data from this database every time an entry is made and to send information related to process control to the process database and information related to scheduling to the scheduling database.

The process database contains information such as which machine group is free to be loaded, which operation was just completed on a machine, which job number, etc.? The presence of a job in this database means that the job is ready to be processed. When the scheduler schedules a job on to the cell, the job which has just been assigned will be removed from the database.

The process planner selects the plan with the highest priority from process data file. The previous operation performed on this machine is known and so the next operation can be generated. In the next step, the process planner considers all the alternative plans created and compares them with the machine database. If a machine specified by the process database is not available, it checks to see if the same oper-

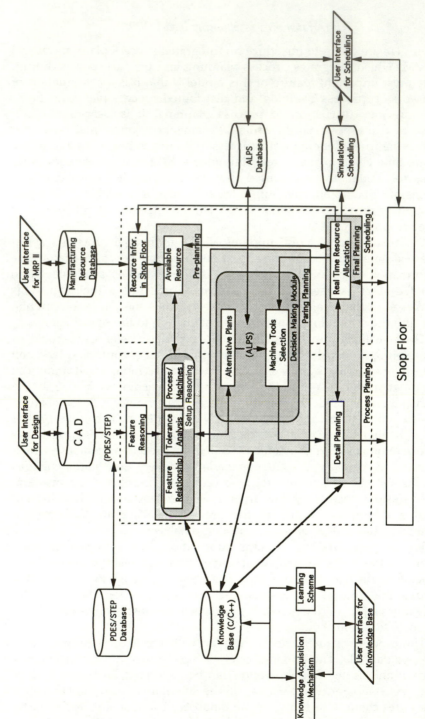

Figure 10.4 The achitecture of IPPM.

ation can be performed by using alternative machines. The necessary information for this is obtained from the machine database. If a particular machine group is obtained, then it generates the parameters required to machine the job on this machine group. Every time a match is obtained, operation time and set up time are calculated and required fixtures are identified. The process planner selects a machine group for every operation and calculates routing information for every alternative. These plans, which are all feasible in principle, are stored in the feasible plans database for later use. It should be pointed out at this point that flexibility is increased by creating alternative plans. The next step is scheduling, and it is evident that if scheduling is to achieve a high degree of efficiency, it must be given a high degree of freedom to take decisions.

The process planner creates plans based on the availability of a machine group, while scheduling assigns a specific operation to a specific machine within a group. To do this, scheduling retrieves information from the schedule database. This database contains information such as which particular machine within a particular machine group is free, what was the previous operation performed on this machine, what was the fixture used etc. It is the responsibility of the process plan to specify what kind of fixture is required. If a new fixture is required, then the previous fixture has to be removed and the required fixture has to be brought from the store or another machine, so the setup time will be higher.

Scheduling takes the alternative plans stored in the feasible plans database and tries to match the requirements with the resources available. It is important that scheduling does not schedule any job on a machine which will go into maintenance at that time. Scheduling assigns a particular job to a particular machine using the Shortest Processing Time (SPT).

The model just described can deal with machine breakdown problems. When a machine breaks down, the job currently being processed on the machine will be blocked and the cell controller reports the status of this job as incomplete. Production control, while interpreting the data, checks if the job is complete; if not, it enters the job number into the process database and the status is reported as incomplete. Process planning gives a high priority to this job. As it is known which machine group failed and which operation was interrupted, the process plan follows the specified steps required to find alternative machines to perform the operation. This information is then sent to the scheduler which follows the specified procedures.

For scheduling and process planning to be integrated, they communicate with each other in some common format. Hence, it is necessary to identify the common information requirement. In IPPM, an attempt has been made to integrate scheduling and process planning using the information available on the shop floor. Emphasis has been given to effective usage of information rather than integration of the two functions based on some mathematical model.

10.7   OTHER ISSUES

In the previous part of this chapter we have discussed a few crucial issues for the development of CAPP systems. The discussion covered software/hardware aspects, user friendly requirements, and integrability. All these issues are important elements for the development of CAPP and have attracted considerable effort. Besides these aspects, there are some other issues which have not been explored very well. Here we think that they are also worth being discussed.

10.7.1   Design support

During the early stage of design, the design engineers usually do not care about such details as a particular bolt pattern, pocket depth, or even the precise shape of a component. Rather, the design engineers care that a part can support desired loads, avoid interfering with other components, and be attached to mating parts. Actually, many current design engineers cannot even make these details appropriately when they are forced to do so. As a result, some poor designs are carried out just because the design engineers do not have enough knowledge about the impact of design variations on the manufacturing process. For example, when a design engineer is going to place a hole in his design part, if there is no influence on the function of the hole, the possible diameters of the hole may cover a range. There will usually be a few standard drills within this range. If the design engineer does not know the standard drill sizes available in his company's tooling department and randomly decides the diameter of the hole, this may cause extra cost and difficulties for manufacturing. Similar problems happen every day in real manufacturing environments. The research emphasis on concurrent/simultaneous (C/S) engineering provides the solution to these problems. Here a process planning based design support system may easily provide information that the design engineer lacks. This system may evaluate the cost and difficulty of manufacturing the design and may suggest specific changes that would reduce production costs. Several research projects are resulting in promising progress. Hayes *et al.* (1989) have developed such a system for design changes suggested by human process planners while creating process plans for machined parts. This work goes through an iterative redesign process of evaluating designs for planar stamped products and making redesign suggestions which will reduce the cost and increase the producibility of the product. The designer can look over the redesign suggestions and decide which one, if any, he wants to use, and the cycle is repeated.

10.7.2   Automated tolerance analysis

Automated tolerance analysis is one of the most critical problems for computer aided process planning systems to be applied in the real manufacturing environment. The traditional way of using CAPP uses only designed shapes and dimensions to generate process plans and operational sequences. Since the tolerance analysis is neglected from this aspect, the process plans are usually infeasible, or at least the plans cannot ensure the dimensional tolerances of the workpiece. Some recent researchers have addressed some of their research activities to computer aided tolerance analysis (Bjorke, 1989; Fainguelernt and Weill, 1986; Tang and Davies, 1988; Lee and Woo, 1989, 1990; Requicha and Chan, 1986; Janakiram *et al.*, 1989; Manivanna *et al.*, 1989; Farmer and Gladman, 1986). All these researchers focused on how to use the computer to assist tolerance analysis, which used to be done manually.

Dimension and tolerance (D&T) chain algorithms are the primary concept for the analysis. The theory of dimension and tolerance chains was originally developed for part assembly because the dimensions of the parts in an assembly are independent. The detailed theory of D&T analysis has been discussed in Chapter 5. The emphasis of how to introduce computer aided tolerance analysis into CAPP systems has been focused on the tolerance chain for operation sequences that are generated based on dimensions and shapes, depending on whether the produced parts are within the designed tolerances. These research efforts use tolerance chains (sometimes referred to as charts) to check operation sequencing. One example was done by Karolin, who introduced computer techniques for solving tolerance chain analysis which used to be done manually. The algorithm used was that which performed the traditional manual graphical method of presenting the dimensions of a workpiece or assembly at all stages of its manufacture (Karolin, 1984). Another example is provided by Li and Zhang, who developed an operational dimensioning and tolerancing calculation in CAPP systems for precision manufacturing. This is a typical checking module. This module is for the calculation of operational dimensions and tolerances by using dimensional chains and graphical methods which can automatically and accurately determine operational dimensions, allowances, and tolerances for all machine operations. Functions of the module include formation of dimensional trees based on input part design dimensions, calculation of operational dimensions and tolerances, and checking the correctness of each final dimension and tolerance against part design dimensions and the allowances for each operation in both lengthwise and diametrical directions (Li and Zhang, 1989). Apart from these two examples, Tang and Davies (1988) and Irani *et al.* (1989) have also discussed some promising results from their research. However, there are still some issues that have not been addressed yet in the area of computer aided tolerance in CAPP systems. For instance, how to select a correct surface as a setup datum for ensuring tolerance requirements from a design is a very important issue for quality issues and feasible process planning. Here, selection of the setup datum will not only involve the analysis of tolerance, but also decision making based on the specific machines to be used. In this case the above check-up approaches are not enough, Because the tolerance analysis has to take place before the operation sequence is generated, the setup position will be used in the analysis.

10.7.3   Process concentration and dispersion

The issue of process concentration and dispersion belongs to the research topic of alternative process planning. Alternative process planning is one of the solutions for integrating process planning with shop floor scheduling, since a process planning system should be able to make all of the alternative plans according to the machine tools available on the shop floor. The system will rank all plans in descending order and submit the first one to shop floor management. If the first plan conflicts with the machine tool schedule, then the second one would be submitted to replace the first. This cycle will be repeated until the optimum machine tool schedule is achieved. The policy of process concentration and dispersion originally belongs to production strategy for different production types. In Chapter 2 we discussed three production types: job shop production, batch production, and mass production. We also discussed that, for a given workpiece, different production types may lead to different production management strategy in terms of processes, machine tools, labor skill level, plant layout, etc. In general, job shop production often requires process concentration, while mass production prefers process dispersion. Process concentration allocates as many processes as possible to one or several machine tools. Since the job shop production type is usually only a single part, less machine tool involvement will reduce the manufacturing lead time and the cost of the product. Contrariwise, mass production will receive great economic benefit by distributing processes over as many machine tools as possible. How to introduce this strategy into CAPP systems and combine the idea of integrating with job shop scheduling has not been addressed very much.

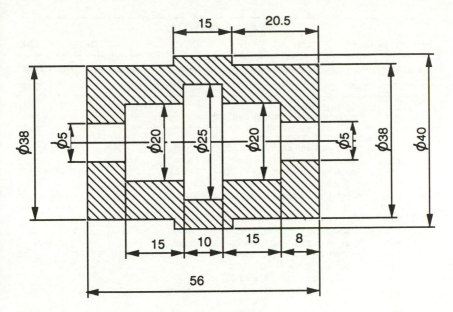

Figure 10.5 A part drawing called EXAMPLE.

10.7.4 Pseudo-process plan recognition

Pseudo-process plans can be defined as plans that are generated theoretically and logically without considering the machinability of a workpiece as related to the limitations of tools, methodology, and technology. For instance, Figure 10.5 shows a rotational workpiece named EXAMPLE. A process plan is carried out and shown in Figure 10.6. Ignoring the limitation of tools, we may think that the process plan is executable. However, there are no effective cutting tools and supports in the current tool base to get through the 5 mm hole to cut and finish the inner surfaces with diameters 20 and 25 mm. In such a case, a pseudo-process plan is generated. It is thus necessary to implement a checking module to identify the machinability of workpieces in CAPP systems. In order to check the pseudo-process plan, a machinability recognition criterion must be established in the knowledge base. Theoretically, the discussion of machinability recognition for components should apply to all technologies, including traditional and modern manufacturing processes. For the example in Figure 10.5, it is impossible for the lathe to machine the inner surfaces, but if precision casting is used the inner surfaces are formed much more easily. How to check pseudo-process plans is a complex problem. It is based on the capacity of the knowledge base of the system and what production rules have been implemented into it. Correction of pseudo-

XPLAN-R PROCESS PLAN

P#	O#	Description		Tools	Setup Tme	Mach. Time
10						
		Raw material storage				
	10	Automate st. bar D=42.00 L=93.42				
20						
		Colchester CNC 2000L				
	10	Clamp part with ext.	L = 78.50	Chuck	.4 h	.09 h
	20	Facing section depth	= 2.50	Facer		
	30	Rough turn diameter	D = 40.00	Sidebit		.74 m
			X = 0.00	L = 56.00		
	40	Rough turn diameter	D = 38.00	Sidebit		.25 m
			X = 0.00	L = 20.50		
	50	Rough turn diameter	D = 38.00	Insertin		.31 m
			X = 35.50	L = 20.50		
	60	R. dri. inte. diam.	D = 5.00	Driller		3.00 m
			X = 0.00	L = 56.00		
	70	R. bor. inte. diam.	D = 5.00			
			X = 0.00	L = 48.00		
	80	R. bor. inte. diam.	D = 20.00	Insertin		
			X = 8.00	L = 15.00		
	90	R. bor. inte. diam.	D = 20.00	insertin		
			X = 33.00	L = 10.00		
	100	R. bor. inte. diam.	D = 25.00	Insertin		
			X = 23.00	L = 10.00		
	110	Fine turn diameter	D = 38.00	Finebit		.51 m
			X = 0.00	L = 20.50		
	120	Fine turn diameter	D = 30.00	Finebit		.51 m
			X = 35.00	L = 56.00		
	130	Cut off the part	1X56.00	Cutter		.40 m
			X = 0.00			

Figure 10.6 Pseudo-plan for EXAMPLE.

process plans is even more important, since this will let the system be used in real manufacturing environments.

10.8 CONCLUSIONS

Computer aided process planning covers a wide area and many technologies have been involved in research and implementation of CAPP systems, as well as in the rapid development of today's computer aided techniques. It is not easy to predict the future trends. In this chapter we discussed several crucial issues about the development of CAPP systems. During the discussion of these issues, the direction of future trends as seen by the authors has been indicated. The need for further efforts on the research and development of CAPP systems is very obvious. Hardware and software selection will still be important as the first step of the development of CAPP. However, as researchers and developers, our attentions should focus on the terms of intelligence and integrability. This is knowledge-intensive in nature. The traditional computer based methods are unable to deal with the challenges of intelligent and integrated process planning because, while they are good at processing data for information-intensive domains, they are not well suited for automatic inferencing for knowledge-intensive tasks. Rather than simply processing information and data, AI-based techniques are designed for capturing, representing, organizing, and utilizing knowledge on computers, and hence will be the key technology for intelligent and integrated process planning in the future. From the discussion of decision making modules, we quickly see that knowledge-based techniques alone are not sufficient to support the requirements for the development of intelligent and integrated process planning systems. A comprehensive and sophisticated decision making module must involve an integrated model of knowledge bases, fuzzy logic, and neural networks. In looking to the future we would like to repeat our earlier sentence again to finish this chapter, 'It is hoped that this book can be considered as "casting a stone to attract jade" – and it offers a few commonplace remarks concerning introduction problems of CAPP so that others may more easily get started' (Alting and Zhang, 1989).

REFERENCES

Alting, L. and Zhang, H.C. (1989) Computer aided process planning: The state-of-the-art survey. *International Journal of Production Research*, **27**,

Bjorke, O. (1989) *Computer-Aided Tolerancing*, ASME Press, New York.

Budde, W. and Zolzer, H. (1973) EXAPT in NC Operation Planning. Proceedings of the 10th Annual Meeting and Technical Conference, NC Society, New York City, April 15–18, pp. 176–90.

Eversheim, W. and Schulz, J. (1985) CIRP Technical Reports, Survey of computer aided process planning systems. *Annals of the CIRP*, **34**(2), pp. 607–11.

Fainguelernt, D. and Weill, R. (1986) Computer aided tolerancing and dimensioning in process planning. *Annals of the CIRP*, **35**(1), pp. 381–6.

Farmer, L.E. and Gladman, C.A. (1986) Tolerance technology – computer-based analysis. *Annals of the CIRP*, **35**(1), p. 7.

Goldberg, A. and Robson, O. (1983) *Smalltalk-80, the Language and its Implication*, Addison-Wesley, Reading, Massachusetts.

Ham, I. and Lu, C.-Y. (1988) Computer-aided process planning: the present and the future. *Annals of the CIRP*, **37**(2), pp. 1–11.

Hayes, C.C., Resa, S. and Wright, P.K. (1989) Using process planning knowledge to make design suggestions concurrently. *ASME-WAM*, DE-Vol. **21**, PED-Vol. **36**, pp. 87–92, San Francisco, CA, December 10–15.

Irani, S.A., Mittal, R.O. and Lehtihet, E.A. (1989) Tolerance chart optimization. *International Journal of Production Research*, **27**(9), pp. 1531–52.

Janakiram, D., Prasad, L.V. and Rao, U.R.K. (1989) Tolerancing of parts using an expert system. *International Journal of Advanced Manufacturing Technology*, (4), pp. 157–67.

Karolin, A.V. (1984) Computer Aided Tolerance Analysis. AUTOFACT 6 Conference, Anaheim, California, MS84-762.

Larsen, N.E. and Alting, L. (1989) Simultaneous Engineering within Process and Production Planning. Proceedings of Pacific Conference on Manufacturing, Australia.

Lee, W.J. and Woo, T.C. (1989) Optimum selection of discrete tolerances. *Journal of Mechanisms, Transmissions, and Automation in Design*, **111**, June, pp. 243–51.

Lee, W.J. and Woo, T.C. (1990) Tolerances: their analysis and synthesis. *Journal of Engineering for Industry*, **112**, pp. 113–21.

Li, J.K. and Zhang, C. (1989) Operational dimensions and tolerances calculation in CAPP system for precision manufacturing. *Annals of the CIRP*, **38**(1), pp. 403–6.

Manivanna, S., Lehtihet, A. and Egbelu, P.J. (1989) A knowledge based system for the specification of manufacturing tolerances. *Journal of Manufacturing Systems*, **8**(2), pp. 153–60.

Nilssen, A. (1989) Workstations still hold the edge. *Mechanical Engineering*, May, pp. 68–71.

Requicha, A.A.G. and Chan, S.C. (1986) Representation of geometric features, tolerances, and attributes in solid models based on constructive geometry. *IEEE Transaction, Journal of Robotics and Automation*, **RA-2**(3), September, pp. 156–65.

Sherif, Y.S. and Heaney, A.A. (1990) Development of a generic database for CAE/CAD/CAM system at an Air Force facility. *Microelectronic Reliability*, **30**(3), pp. 555–64.

Tang, X. and Davies, B.J. (1988) Computer aided dimensional planning. *International Journal of Production Research*, **26**(2), pp. 283–97.

CHAPTER 11

Building a CAPP system – a case study

11.1 INTRODUCTION

Most of the fundamental technologies related to the subject of process planning have been discussed in previous chapters, such as GT, CAD, CAM, concepts of CIM, some approaches of CAPP, application of KBES in PP, as well as the techniques involved in the implementation of CAPP systems. All these form a part of the objectives of this book. In this chapter we are going to demonstrate how to build a CAPP system by means of a case study – an expert process planning system focused upon rotational components. The system has been constructed and is entitled XPLAN-R. XPLAN-R was developed at the Laboratory of Process and Production Engineering in the Institute of Manufacturing Engineering (IME) at the Technical University of Denmark (TUD). The system was based on XPLAN, which is a general framework model for process planning created by Lenau in 1984 (Lenau and Alting, 1986) and further developed by Zhang from 1986 to date (Zhang, 1987), (Alting *et al.*, 1988). This development has provided a hybrid approach for knowledge representation by combining the rule-based and frame-based approaches. XPLAN-R is based on an expert system building environment, the DCLASS tree processor, and is being interfaced with a relational database, R:BASE System V. Some general information about both DCLASS and R:BASE System V will be briefly introduced in the following sections. The architecture of the system and the structure of the DCLASS database will then be presented in section 11.5. Currently, the DCLASS database contains almost all of the important data while the system is being used. Thus far, R:BASE System V is being interfaced with XPLAN-R as a partial database of XPLAN-R. Some of the primary databases have been established based on R:BASE System V, such as the raw materials database, machining database, cutting tools database, and so on. It is necessary to introduce the structures of these databases implemented in the format of R:BASE System V. Following this, the information flow and work steps of XPLAN-R will be presented in section 11.7. The information flow indicates how the data and knowledge (decision-rules) are processed while the system is being executed. At the present time, XPLAN-R works in nine steps, as follows:

1. part specification
2. process selection
3. stock determination
4. machine selection
5. fixtures selection
6. operation sequencing
7. tools sequencing
8. cutting condition decision
9. time and cost calculation

These nine work steps will be introduced in detail in sections 11.7.1–11.7.9. On the basis of the above discussion, three examples will be tested in section 11.8. These three examples belong to three fuzzy boundary part families: the screw family, the shaft family, and the flange family. The detailed process plans corresponding to each of these examples will be presented in the latter section.

11.2   XPLAN-R: GENERAL VIEW

XPLAN-R stands for eXpert process PLANning system for Rotational components. It is based on an expert system building environment, DCLASS, which is a standard (hierarchical) classification tree processor. XPLAN-R uses the DCLASS decision tree processor in building up its knowledge base and the partial database which contains the general process planning decision rules and the data being processed. R:BASE V is a relational database product which can be used for designing a set of tables to store and retrieve data easily (R:BASE System V, 1986). Some relational sub-databases have been established in the form of R:BASE System V tables, such as the raw materials sub-database, the machining sub-database, and the cutting tools sub-database, etc. The input data necessary for XPLAN-R is entered into the system by means of a user-friendly and highly interactive session. The system allows for an easy dialogue between the user and the system. Since XPLAN-R already contains most of the decision logic necessary for rotational machining, the user does not need a detailed manufacturing technology background. He/she only needs to answer a series of questions about the part specifications, such as batch size, material properties, dimensions, tolerances, roughness, surface treatments, geometrical features, etc. Most of the answers can be found on a standard machining blueprint. The questions are either of the multiple choice type or variable value type. The output data from the system can be printed out as an individual process plan sheet or an individual operation plan sheet, which are well suited for shop floors. A material bill, a machine tools sheet, and fixturing and tooling sheets can also be individually printed out from R:BASE V. All these output data can be interfaced easily with other systems, such as an MRP system, a production simulation system, and a CNC machine or an MRP II systems.

11.3 DCLASS TREE PROCESSOR

As we discussed in Chapter 4, the DCLASS system and technology was conceived in 1976 by Allen and Millett at Brigham Young University (DCLASS Technical Manual, 1985). In 1978, the first application and installation was made at Boeing Commercial Airplane Co. for the purpose of generative planning for aircraft parts. Since then, the system has grown to become a powerful, compact and efficient information handling and decision making system capable of being tailored to fit each user's needs.

DCLASS is an acronym for Decision CLASSification Information System. It is a general purpose computer system for processing classification and decision-making logic. The system has two main features:

- Information tree processing that allows both standard and user defined logic
- Flexibility which allows easy interfacing with the user's own application program environment

The tree structures used in DCLASS may contain classification systems or other user-defined logic. As we discussed in Chapter 4, DCLASS has been very well used in the application of GT, especially for classification and coding systems.

Classification

Classification or coding schemes may be readily formatted into DCLASS trees. Once in the tree structure, these classification systems are easy to update and maintain. Desired modification to the classification trees is easily accomplished.

Coding

With DCLASS, the code length is flexible. It can accept any number of digits depending upon the needs of the application. Various parts of the code may be used to provide pointers into specific parts of the database or they can be appended to make a comprehensive code of any desired length. These codes can be fixed length monocodes or polycodes or more flexible variable length codes.

Information retrieval

Information retrieval with DCLASS technology is fast. The degree of match between the defined target item and what is currently in the database may be varied from a perfect match to any user specified degree of similarity. DCLASS retrieval is not limited to searching and sorting on a fixed length code.

In this chapter, our discussion will emphasize how DCLASS can be used in conjunction with existing database systems for storage of codes and variables and for subsequent retrieval using a traditional database management system. This feature is used for interfacing XPLAN-R with R:BASE V.

Decision making

DCLASS provides a very simple and straightforward method of constructing decision making logic trees. DCLASS trees contain the logic, sequences, calculations, keys, data elements, and codes used for such diverse activities as Generative Process Planning (GPP), Automated Time Standards (ATS), N/C and Robotics Programming, Automatic Materials Selection (AMS), and even Parametric Production Design. Decision trees permit the user to relate conditions and actions. This capability performs an essential function in XPLAN-R.

System integration

One of the very useful benefits of the DCLASS technology is its ability to integrate quite diverse CAD and CAM application programs. Any node of the DCLASS system can be used to issue a subroutine call and pass data between various application programs.

DCLASS can process various trees for classification, coding, or decision making and then pass the resulting codes and values to the mainline control module for use with other application programs.

ES building environment

Since highly automated tree processing capabilities are available in addition to multiple-path and multiple-level branching capabilities, DCLASS may be best defined as a tree processing system with a very high level tree definition language (DCLASS Technical Manual, 1985). This is why DCLASS can not only be used to derive codes, but can also handle complex decision making and 'expert' AI problems. According

```
TE SU LE MX PA OC  PR  CO IN OU DE CM CH FI LO NU PA EX XR BI LI FO  4.0
30 15 40 99  8  1 132   1  1  1  0  1  0  0  1  0  1  0  0  0  0  1

DCLASS TREEDRAW:  ** ZY   ** Job 1   8/ 2/1989  14:29:24         Page  1


                              <IF> Thread      <THEN> threading process
    For attributes on a    :---------------:  is required
    diameter               %:                 :---------------------------
    -------------ZH----------:
                           :  <IF> Keyway      <THEN> milling process is
                           :---------------:   required
                                              :---------------------------

1
```

Figure 11.1   An example of IF-THEN performance in a DCLASS tree.

to the definition of ES building environments discussed in section 3.5, DCLASS, as a very high level programming language, can be classified as an ES toolkit. Seen as such, DCLASS can be used to analyze and capture production rules, decision making logic, and technical know-how for a specific production environment or from experienced process planners. The expertise can be easily and consistently used by others. The knowledge representation in DCLASS is a typical rule-based approach, as discussed in section 3.5. That means an IF-THEN structure is performed in the form of a tree. A simple example is shown in Figure 11.1.

Some general information is very briefly listed here. For a further description of the use of DCLASS, the DCLASS user's manual may be studied.

DCLASS can be chosen as either a small DCLASS system or a large DCLASS system. Their specifications are:

1. The small system allows definition of 676 sub-trees.
2. The large system allows definition of 4080 sub-trees.
3. Last 20 ID#s stored may be displayed and/or listed without leaving DCLASS and running DCUPDAT.
4. Each sub-tree can maintain 150 branches.
5. Each sub-tree can call 25 other sub-trees and 20 recursive calls are allowed for itself.
6. Local and Universal file sizes up to 32768 records.
7. 16 child branches per parent branch.
8. 60 main sub-trees.
9. Approximately 246 characters of text per branch, with a total text limit of 4000 characters per sub-tree (including input and output keys, and codes which are greater than 10 characters).
10. 1000 variables for each ID#.

11. 1000 output keys (path names and branch flags) for each ID#.
12. 3200 characters of code (the first 128 being retrievable by the existing DCLASS mainline).
13. 1000 bit string size.

11.4   R:BASE SYSTEM V

R:BASE System V is a relational database product which can be used for designing a set of tables to store and retrieve data easily. An R:BASE database is a collection of up to 80 of these tables. Each R:BASE database may have a combined total of up to 800 columns for all tables. For each column, a data type tells R:BASE what kind of data the column will hold – such as codes, text, dates, variables, etc. Columns can also hold a value computed from other columns.

The power of R:BASE System V is in the control it gives users over the tables they create. The data from one or any combination of the tables within a database can be manipulated. Tables can be organized so that any particular piece of information rarely needs to be entered more than once. R:BASE System V allows users to put together information from the tables so that the combined information answers the user's questions. Further details and a description of R:BASE V can be found in the User's Manual of R:BASE System V (R:BASE System V, 1986).

11.5   THE SYSTEM ARCHITECTURE AND DCLASS DATABASE

The architecture of XPLAN-R is illustrated in Figure 11.2. It indicates that XPLAN-R, so far, works from the specifications to the process plan in nine steps. Each step is controlled and communicated between the knowledge base and the databases. The knowledge base, which contains the knowledge of process planning and the know-how of the technology for manufacturing rotational components, is built up by means of the expert system's tool kits – the DCLASS decision tree processor. The typical rule-based approach for knowledge representation in the form of IF ⟨antecedent⟩ THEN ⟨consequence⟩ is performed by decision sub-trees. The detailed processing of knowledge representation is indicated in the following sub-sections, which give a description of each work step. The global architecture of the XPLAN-R system is built upon the standard hierarchical classification sub-trees.

The database of the system consists of two components, the DCLASS database and the R:BASE System V database. At the moment, as a built-in database system within its programs, the DCLASS database takes the dominant functions. In XPLAN-R, the DCLASS database

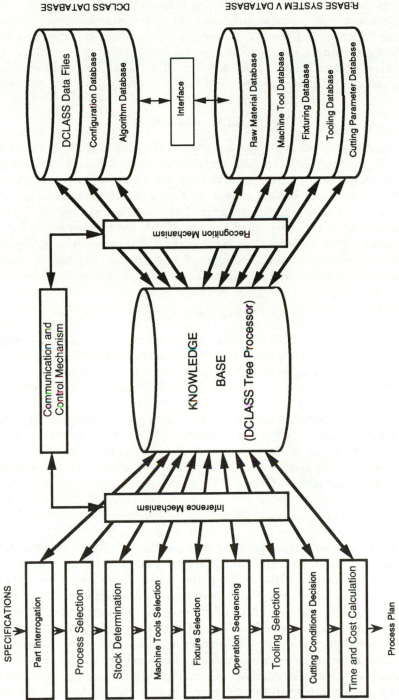

Figure 11.2 The architecture of XPLAN-R.

is divided into several sub-databases, namely DCLASS data files, a configuration sub-database, and an algorithm sub-database, while R:BASE V consists of the raw materials sub-database, machining sub-database, operation sub-database, tooling and fixturing sub-database, and DCLASS data files. The functions of all of these sub-databases will be introduced as follows:

Configuration sub-database

The configuration sub-database is used to store and to retrieve the data of the geometric shape of the workpiece planned by XPLAN-R. A pattern recognition mechanism has been built in and is stored with a combination of several algorithms in the algorithm sub-database to distinguish the shape of the workpiece. Meanwhile a strategy of fuzzy boundary part families has been embedded into the pattern recognition mechanism. When a workpiece is introduced into the system, it will be automatically recognized according to its geometrical configuration and stored within the relevant family within the sub-database.

Raw materials sub-database

The raw materials sub-database is used to store and retrieve the data for raw materials corresponding to the workpiece planned by XPLAN-R. Usually, three groups of data can be provided here. They are: (1) the specific physical properties of materials such as carbon steel, stainless steel, nylon, etc.; (2) the shape of raw materials such as long bar, hexagon bar, square section, etc.; and (3) the dimensions of the raw materials such as lengths and diameters. When a workpiece is put into the system, the corresponding data for the raw material will be matched and retrieved from the sub-database according to the data specification of the workpiece.

Machining sub-database

The machining sub-database is used to store and retrieve the data for the machines used to carry out the selected manufacturing processes. Currently, machining data is stored and retrieved in the form of ID#s in XPLAN-R, where each ID# corresponds to a specific machine. The data for the ID#s are stored in the machining sub-database. The sub-database contains the specifications of machines within a production environment, for example a company, a workshop, or a production cell. The data include items such as the function of the machines (milling or turning or milling center, workstation, etc.), the main dimensions of the machine (X-axis, Y-axis, Z-axis), tool positions, power, speed arrangements, etc.

Operation sub-database

The operation sub-database is used to store and retrieve the data for operation sequencing planned by XPLAN-R. This sub-database is associated closely with the knowledge base of the system. The data is frequently retrieved and stored by the system; then the results are inferred to decide the appropriate sequence of operations. Some cutting conditions are stored in this sub-database if the system is employed in real production environments.

Tooling and fixturing sub-database

The tooling and fixturing sub-database is used to store and retrieve the descriptions of tools and fixtures corresponding to the machines stored in the machining sub-database. At present, this sub-database is simple, holding only a few data about tools, such as sidebit cutter, insert cutter, driller, finished cutter, etc., and some simple fixtures or jigs, such as three-jaw chuck, center-driller, tailstock supporter, etc. In order to provide precise and detailed data more effectively about tools and fixtures, a proposal for using R:BASE System V to store and retrieve the data instead of DCLASS sub-tree structures has been carried out. This idea will be introduced in section 11.7.

Algorithm sub-database

The algorithm sub-database is used to store and retrieve all necessary calculation programs used in XPLAN-R. When XPLAN-R is executing, some calculation programs are called and employed to perform calculations such as sorting diameters, deriving parameters, calculating machining time, etc. These calculation programs are written as either DCLASS sub-trees or FORTRAN subroutines and are stored in this sub-database.

DCLASS data files

The DCLASS data files are the primary data files for storing all necessary information in XPLAN-R, such as the configuration of sub-trees, codes, ID#s, local keywords, etc. In general these data files are divided into two major groups; one is the main set of files that is accessed by multiple users, the other includes the specific update and temporary files for a single user. Each group consists of several specific data files, all of which can be accessed and updated very easily by executing DCLASS commands. For instance, DCPREP is for updating new subtree configurations (new branches, code or keywords), DCUPDAT is for updating new ID#s, etc. Thus, this sub-database functions as a Database Management System (DBMS) in XPLAN-R.

Since the main features of the system are in the form of DCLASS tree structures, all of the necessary data (variables, codes, keywords, etc.) are retrieved and updated in the DCLASS database. However, when the databases are filled with considerable data, the tree structures become larger and larger. This causes the execution time to be increased for complex workpieces. In order to resolve this problem, a proposal for interfacing XPLAN-R with R:BASE System V has been carried out. The basic concept of the proposal is to develop some R:BASE System V databases instead of the DCLASS database. R:BASE System V is a standard relational database system which stores and retrieves data in the form of tables. Thus, it is easier and faster from the point of view of processing large amounts of data. Other benefits also result from this proposal; for instance, interfacing R:BASE V will provide the possibility of utilizing XPLAN-R in multiple production planning rather than process planning only. A brief introduction to these benefits will be given in section 11.8. Several databases in R:BASE System V format have been established in XPLAN-R.

11.6   R:BASE SYSTEM V DATABASE

As mentioned previously, R:BASE System V is a relational database system which can be used for designing a set of tables to store and to retrieve data easily. Some prototypes of these databases have been established. The fundamental relationships between these prototype databases is illustrated in Figure 11.3.

Five databases have been designed in R:BASE System V. They are the raw materials database, the machine database, the tooling database, the fixturing database, and the cutting parameters database. Figure 11.3 not only illustrates the relationships between these databases, but also indicates how the data is represented in R:BASE V. From Figure 11.3 one can see that the relationships between these databases are based on a network structure. It indicates that some databases are related to other databases. For example, the fixturing database is related to the machine database, which means that fixturing data stored in the fixturing database must match data in the machine database. Each machine has its own fixtures stored in the fixturing database. Furthermore, the tooling database is related to both the raw materials database and the machine database. This means that the data for tools depends on the data for both the materials to be cut and the machines to be employed. As shown in Figure 11.3, some databases consist of one set of tables, such as a machine database, and some databases consist of two sets of tables, such as the raw materials database and the tooling database. This indicates how the data is to be represented in the database. If the database consists of one set of

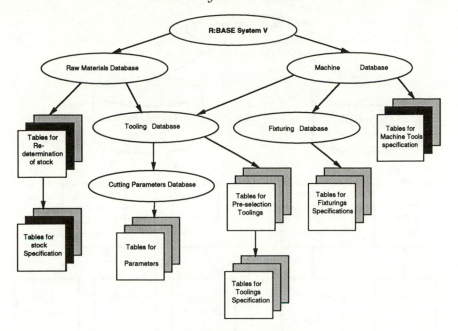

Figure 11.3 The fundamental relationship of R:BASE V databases.

tables, the data will be represented directly from columns in the tables. However, if a database consists of two or more sets of tables, a set of intermediate codes will be used to transfer the necessary precise data. In Figure 11.3, some of the tables are shown with dark backgrounds, indicating that these tables have already been established, while other tables are shown with light backgrounds, indicating that these tables have not yet been established. Thus far, two databases with several sets of tables have been developed within the thesis work. Figure 11.4 illustrates the current attributes of R:BASE V databases in XPLAN-R.

11.6.1   Database MACHDATA

MACHDATA is a database which contains machine specifications. A set of tables has been implemented. They are table (1), entitled **Turning**, which contains 10 CNC lathes, and Table (2), entitled **Milling**, which contains 7 CNC milling centers. Each table has several columns which contain the specifications for each machine, as shown in Figure 11.5. They are printed out directly from R:BASE System V. Tables can be printed out in many different ways, and there are many columns to define the machines. Due to the limited width of the paper, only four columns are printed out here. They are (1) ID#s of the machines, (2) the names of machines, (3) the types of machines, and (4) the group

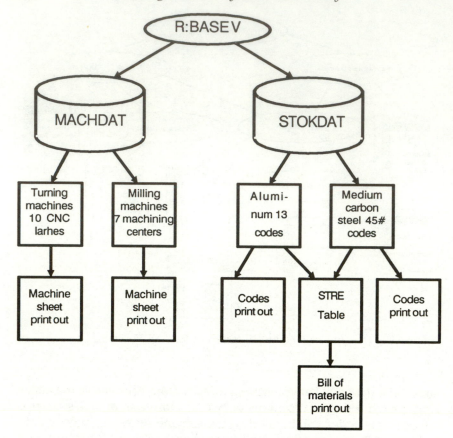

Figure 11.4   Current attributes of R:BASE V databases in XPLAN-R.

specification codes. Figure 11.6 shows that machine sheets can be printed out from R:BASE V. Each row in the tables (Turning and Milling) stands for one specific machine. When a machine is selected by XPLAN-R, the sheet will be printed out from R:BASE V. Figure 11.6 is an example of a machine sheet. It is a CNC lathe named OKUMA LB15-1SC 500. The machine's ID# is TU08 implemented in XPLAN-R. If desired, more information can be easily input into the tables by creating more columns.

11.6.2   Database STOKDATA

STOKDATA is a database which concerns stock specifications. Two sets of tables have been implemented to represent the final stock specifications determined by the system; they are named STPD (STock, Pre-Determination) and STRE (STock REpresentation). The STPD con-

```
machi  machname           machtype      spnbo  spnspeed  motpower          costmi
-----  ----------------   -----------   -----  --------  ----------------  ------
TU01   BOEHRINGER VDF     CNC LATHE     130.   14-2240   60KW              300.
       400C
TU02   BOXFORD 300-IS     NC LATHE       35.   27-3000   1.5KW             180.
TU03   COLCHESTER CNC     CNC LATHE      54.   20-2750   5KW (7.5HP)       250.
       2000L
TU04   CORTINI H105       CNC LATHE      20.   150-3000  1HP               205.
TU05   DYNA MYTE 3000     CNC LATHE      16.   0-4000    3/4 HP Servo      215.
                                                         motor
TU06   MORI SEIKI CNC     CNC LATHE      77.   30-3000   DC22/30KW         270.
       SL-5
TU07   NAKAMURA SUPER     CNC LATHE      80.   10-3500   AC 11/15KW        280.
       TURN 2B
TU08   OKUMA LB15-1SC     CNC LATHE      56.   75-4200   11KW              290.
       500
TU09   PARTNER L-100H     CNC LATHE      32.   80-4000   AC 5.5KW          210.
TU10   YAMAZAKI QUICK     CNC LATHE      80.   13-3000   25HP/18.7KW       270.
       TURN 20


       * * * * * * * * * * * * * * * * * * * * * * * * * * * * * * *


machi  machname           machdife              tolp.  spnspeed   motpower
-----  ------------------  -------------------  -----  ---------  ---------------
ML01   CORTINI L300       CNC VERTICAL MINI     6      150-3200   0.3KW(1.5HP)
                          MACHINING CENTER
ML02   FANUC TAPE         CNC MINI              10     35-4500    7.5 KW
       CENTER-MODEL D     MACHINING CENTER
ML03   HERMLE UWF 700     CNC MILLING AND       1      40-2000    2.2KW
       CNC                DRILLING MACHINE
ML04   MAZAK VQC-20/40    CNC DOUBLE-COLUMN     20     28-4000    AC7.5HP/18KVA
                          MACHINING CENTER
ML05   MITSUI SEIKI HR3B  CNC                   30     45-4500    11/15 KW.AC
                          HORIZONTAL-SPINDEL
                          MACHINING CENTER
ML06   OKUMA MC-4VA       CNC VERTICAL          20     50-4000    VAC3.7/5.5 kw
                          MACHINING CENTER
ML07   PARTNER M-300      CNC VERTICAL          20     100-6000   AC 3.7/2.2 KW
                          MACHINING CENTER
```

Figure 11.5 Ten CNC lathes and seven CNC milling centers implemented into the machine tools database.

```
                    ***********************
                    *   MACHINE SHEET     *
                    ***********************

            Institute of Manufacturing Engineering
               Technical University of Denmark
                  DK-2800 Lyngby Denmark

    ----------------------------------------------------------
          This machine titled: OKUMA LB15-1SC 500
    is selected by the expert process planning system XPLAN-R
                    Date: 08/02/89
    ----------------------------------------------------------

                    MACHINE SPECIFICATIONS
                    ----------------------

    Max. swing over bed:   400mm    Max. machining diameter:   250mm
    Max. longit. travel:   520mm    Max. machining length:     470mm

    Bar work machining:  56mm       Tool positions: 12
    Spindle speeds: 75-4200  r.p.m.
    Main motor power: 11KW

                    MACHINE CLASSIFICATIONS
                    -----------------------

    Machine ID number: TU08         Type: CNC LATHE
    Cost per. minute:    290.DKR     Function code: DN
    Production cell location: MLP.SUP-DP(N) "LM"
    Installed date: 01/10/89
    ----------------------------------------------------------
```

Figure 11.6 An example of a machine sheet.

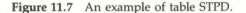

```
********************************
       REPORT FORM STPD TABLE
********************************
                08/02/89

Condit:              20.00mm   25.00mm   30.00mm   35.00mm   40.00mm   45.00mm
-----------------    --------  --------  --------  --------  --------  --------
Black bar            BLM45020  BLM45025  BLM45030  BLM45035  BLM45040  BLM45045
Bright bar           BRM45020  BRM45025  BRM45030  BRM45035  BRM45040  BRM45045
Hexagonal bar        HXM45020  HXM45025  HXM45035  HXM45035  HXM45040  HXM45045
Squaredsection bar   SQM45020  SQM45025  SQM45030  SQM45035  SQM45040  SQM45045
```

Figure 11.7   An example of table STPD.

```
              S T O C K   I N V O I C E            DATE: 08/02/89

                  * * * * * * * * * * * * *

            Institute of Manufacturing Engineering
               Technical University of Denamrk
                   DK-2900 Lyngby, Denamrk

        -------------------------------------------------------
            This STOCK INVOICE is used for the RWA MATERIALS DATABASE
            of the expert process planning system XPLAN-R
        -------------------------------------------------------

Property of stock: Medium carbon steel 45#
                   ----------------------------

Codes      Features       Dia.(mm) Length(m) Cost/m(DKK) Stock state Location
--------   ------------   -------- --------- ----------- ----------- --------
HXM45040   Hexagonal bar    40.      10.        285.      Available   MC.04.8
HXM45045   Hexagonal bar    45.      10.        285.      -0-         MC.04.9
SQM45020   Squaredsection   20.      5.         250.      Available   MC.05.3
SQM45025   Squaredsection   25.      5.         280.      -0-         MC.05.4
SQM45030   Squaredsection   30.      5.         285.      Available   MC.05.5
SQM45035   Squaredsection   35.      -0-        -0-       -0-         -0-
SQM45040   Squaredsection   40.      5.         320.      -0-         MC.05.6
SQM45045   Squaredsection   45.      -0-        -0-       -0-         -0-
BLM45020   Black bar        20.      8.         100.      Available   MC.01.4
BLM45025   Black bar        25.      8.         100.      Available   MC.01.5
BLM45030   Black bar        30.      8.5        110.      Available   MC.01.6
BLM45035   Black bar        35.      9.         110.      Available   MC.01.7
BLM45040   Black bar        40.      10.        120.      Available   MC.01.8
BLM45045   Black bar        45.      10.        120.      Available   MC.01.9
BRM45020   Bright bar       20.      10.        150.      Available   MC.02.4
BRM45025   bright bar       25.      10.        160.      Available   MC.02.5
BRM45030   Bright bar       30.      10.        165.      Available   MC.02.6
BRM45035   Bright bar       35.      10.        170.      Available   MC.02.7
BRM45040   Bright bar       40.      10.        175.      Available   MC.02.8
BRM45045   Bright bar       45.      10.        180.      -0-         MC.02.9
HXM45020   Hexagonal bar    20.      8.         250.      Available   MC.03.4
HXM45025   Hexagonal bar    25.      8.         255.      -0-         MC.03.5
HXM45030   Hexagonal bar    30.      8.         265.      Available   MC.03.6
HXM45035   Hexagonal bar    35.      10.        275.      -0-         MC.04.7
```

Figure 11.8   An example of tables in the set STRE.

sists of tables for stock pre-determination. Each table corresponds to specific raw materials. Two prototype tables have been implemented in STPD, one for Aluminum 13# and the other for Medium Carbon Steel 45#. The first column of each table is used to describe the specific attributes of each raw material stored in the database, such as black bar, bright bar, hexagonal bar, squared section, etc. The first row of the

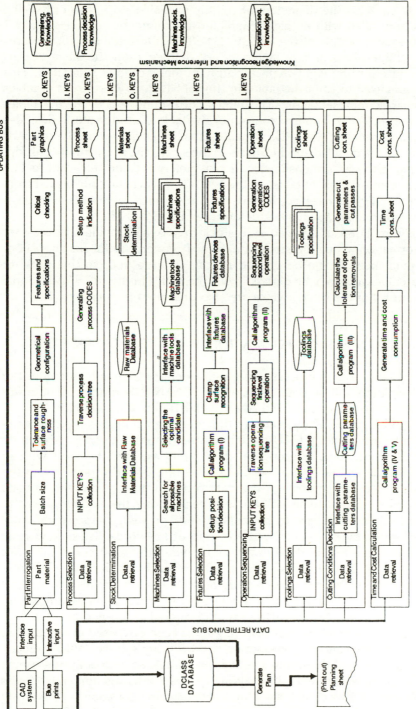

Figure 11.9 The information flow diagram of XPLAN-R.

table contains the diameters of readily available stock; the rest of the space is filled with an eight-character alphanumeric code. Figure 11.7 shows an example of the table in set STPD. Once the specific features and dimensions of raw materials are defined from the stage of part specification, an eight-character alphanumeric code is generated. This code is used to enter the tables in the set STRE. STRE consists of tables for stock representation. The tables correspond exactly to the tables in the set STPD. The first column of the table is used to store the codes corresponding to the codes issued by the STPD; the rest of the columns contain the detailed specifications of the stock, such as attribute descriptions, diameter dimensions, cost per meter, location, etc. Figure 11.8 shows an example of tables in the set STRE. Once the code issued from the set STPD is received, it leads to a row which represents all necessary information about stock determined by XPLAN-R.

Figure 11.8 can be used as a stock invoice to provide all necessary information about the stock situation. It is very useful for shop floor management. It can be printed in any style desired, so that a bill of materials can easily be derived from the tables in the STRE. Additional examples of the stock invoice are also presented in Appendix A.

11.7 SYSTEM INFORMATION FLOW AND WORK STEPS

As it is an expert process planning system, a tremendous amount of information will be dealt with while XPLAN-R is running. The information flow diagram of XPLAN-R is illustrated in Figure 11.9.

From Figure 11.9 one can see that the original data is input interactively, from either a CAD model or a machining blueprint. According to the work steps, the original data input will then be dealt with in steps, one by one. At the start of each step, data is retrieved from the DCLASS database through the data retrieval bus. At the end of each step the updated data will be sent back, using the updating bus, to the DCLASS database to be stored. The knowledge representation is performed by means of input and output keywords. These keywords link the sub-trees of the knowledge base and the sub-trees of the work steps. The final process plans can be printed out from the DCLASS database. At the end of each work step, the individual results, such as a process sheet, a machines sheet, a bill of materials, a tooling sheet, and a time and cost calculation report, should be printed out.

The activities of process planning have been discussed in section 2.5. The primary concept in designing the work steps in XPLAN-R is based on this discussion. In its present form, XPLAN-R works in nine steps, as shown in Figure 11.2. They are (1) part interrogation, (2) process selection, (3) stock determination, (4) machine selection, (5) fixturing selection, (6) operation sequencing, (7) tooling selection, (8) cutting conditions decision, and (9) time and cost calculation. All these work steps will be introduced in the coming sub-sections. Some flowcharts for these work steps are illustrated in Appendix B.

11.7.1 Part interrogation

Part interrogation consists of two main elements: raw material specification and geometrical configuration specification. According to the requirements of the design, the characteristics of materials such as metal or plastic, steel or aluminium, carbon steel or stainless steel, etc., are defined by traversing the classification tree. The conditions of materials such as bar shape, rod, hexagon, etc., are also defined in the same way. The geometrical configuration specification is a major task in this step. In order to describe a complicated rotational shape with a variety of features, a series of questions have to be answered. The configuration of diameters are specified one by one. For instance, for diameter no. 1 (the order of the diameter numbers is from left to right), the questions start from the main shape such as curved or straight, cylindrical or tapered, dimensions, roughness and tolerance, as well as features on the diameter such as chamfers, grooves, keyways, threads, etc. When all of these geometrical attributes have been specified, the system goes to the next diameter. Here, a construction of pyramid-like networks of know-how is employed to build up the classification trees. The main taxonomy for part specification is given in Figure 11.10 to illustrate this concept. In addition, XPLAN-R asks only the relevant questions. If, for instance, a concave diameter is specified by the user, then the chamfers, threads, knurls, etc., which obviously cannot be machined on the concave diameter, will never be considered by XPLAN-R.

XPLAN-R offers debugging help in the TREEDRAW facility. This simply draws the tree structures so that it is easy to get a good view of the tree structure. As an example, a sub-tree which corresponds to the sixth layer of Figure 11.10, where it is marked with a dark circle, is illustrated in Figure 11.11, which shows the possible attribute features for each diameter. When the features are selected according to the design requirements (for instance, a chamfer option and a knurl option) the OUTPUT KEYS will then be defined. The OUTPUT KEYS, here [CHAMF] and [KNURL], are bracketed thus [ ]. These OUTPUT KEYS are associated with the INPUT KEYS, here (CHAMF) and (KNURL) which are parenthesized thus ( ), in the operation sequencing sub-tree. They constitute decision rules which allow the corresponding operation sequencing branches to be selected and traversed automatically. As the chamfer and knurl are defined, further questions about the chamfer, such as left or right chamfer, and the knurl, such as diamond or straight knurl, are asked. The associated dimensions for each feature will finally

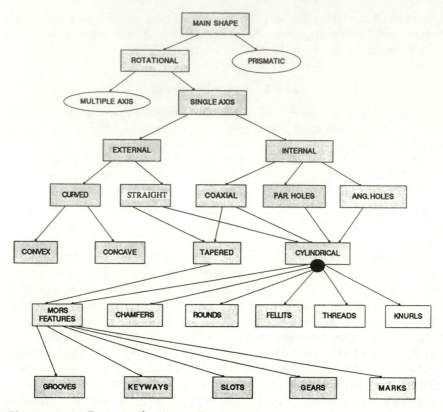

Figure 11.10 Part specification main taxonomy.

be asked. After answering all of these necessary questions, the input data on the geometrical configuration will be transformed and stored in the configuration database for use in the next working step.

As mentioned previously, the idea of 'fuzzy boundary part families' has been introduced in this work step, based on a GT concept. The difference here is that part families are not distinguished by coding systems but by a built-in pattern recognition mechanism. The boundaries between part families are fuzzy, meaning that a critical part might be classified into either of two part families. In spite of slight differences, an identical part classified into either of two part families will have the same or very similar final process plans. The biggest advantage of this idea is to introduce the GT concept into the generative process planning system without the use of coding systems. Since the workpieces can be grouped into part families, the manufacturing processes, shop floor arrangements, and equipment can also be grouped according to the main features of the part families. This idea will enable an easy introduction of the GT concept and the expert process planning

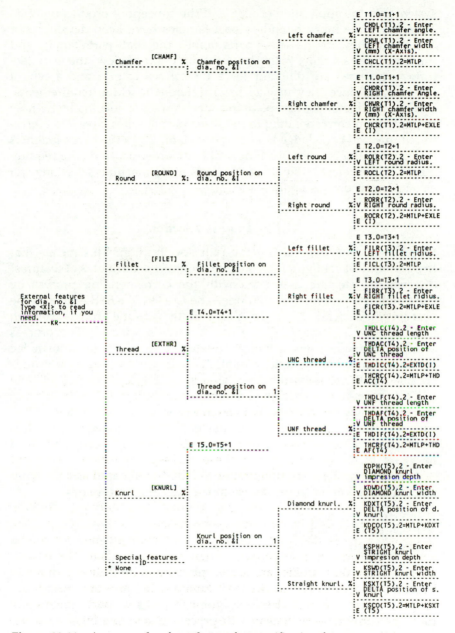

Figure 11.11 An example of a sub-tree for specification features.

system into the application of FMS and the concept of production cells. Thus far, three fuzzy boundary part families have been developed in XPLAN-R; they are the screw parts family, the shaft parts family, and the flange parts family. From the optimum point of view, the manufacturing processes are different for a typical screw part and a typical flange part, since they usually have different machine requirements. The knowledge of process planning, decision logic and the know-how of production corresponding to the three part families have been introduced into the knowledge base of the system. Once the part's family is recognized, the process planning will be oriented towards the requirements of the relevant family, and all resulting machine selections will also be based on this family.

11.7.2   Process selection

Process selection depends on the part specification. This means that the geometrical configuration, surface roughness, attributes of features, dimensions, etc. are associated with the corresponding process by traversing the process selection tree. The OUTPUT KEYS issued in the part specification are compared automatically with the process selection tree. If there is a match the related branch is selected. The process selection is made using a typical rule-based approach for knowledge representation. For example, a rotational part with a keyway feature is specified in the classification tree. Then the OUTPUT KEYS [ROT] and [KEYS] are generated which will match the INPUT KEYS (ROT) and (KEYS) so that the turning and milling processes are selected. When each process is selected, a two-character CODE is generated which is used for machine selection. The process selection sub-tree is illustrated in Figure 11.12.

According to the primary theory of manufacturing engineering, types of production can be generally classified into three categories:

- Job shop production
- Batch size production
- Mass production

In principle, the different types of production will imply different process selections for an identical workpiece. Job shop production and batch size production usually point to process concentration and mass production usually points to process dispersion. This idea has also been introduced into XPLAN-R. The types of production are simplified into two categories, either batch size production or quantity production. In XPLAN-R, batch size production is oriented towards process concentration and quantity production is oriented towards process dispersion. A criterion is defined here to distinguish the two types of production. This result will affect the following work step, i.e.

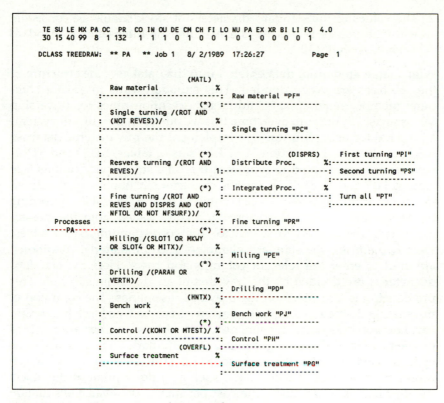

```
TE SU LE MX PA OC  PR  CO IN OU DE CM CH FI LO NU PA EX XR BI LI FO  4.0
30 15 40 99  8  1 132   1  1  1  0  1  0  0  1  0  1  0  0  0  0  1

DCLASS TREEDRAW:  ** PA   ** Job 1    8/ 2/1989  17:26:27          Page  1


                                   (MATL)
                            Raw material           %
                    :-----------------------------:  Raw material "PF"
                    :                         (*)  :-----------------------
                    :  Single turning /(ROT AND
                    :  (NOT REVES))/                %
                    :-----------------------------:  Single turning "PC"
                    :                                :-----------------------
                    :
                    :                         (*)              (DISPRS)   First turning "PI"
                    :  Resvers turning /(ROT AND   Distribute Proc.  %:--------------------
                    :  REVES)/                    1:-----------------------:  Second turning "PS"
                    :-----------------------------:                         :---------------------
                    :                         (*)  :  Integrated Proc.      %
                    :  Fine turning /(ROT AND       :-----------------------:  Turn all "PT"
                    :  REVES AND DISPRS AND (NOT                              :---------------------
                    :  NFTOL OR NOT NFSURF))/     %
        Processes   :-----------------------------:  Fine turning "PR"
        ----PA----: :                         (*)  :-----------------------
                    :  Milling /(SLOT1 OR MKWY
                    :  OR SLOT4 OR MITX)/          %
                    :-----------------------------:  Milling "PE"
                    :                         (*)  :-----------------------
                    :  Drilling /(PARAH OR
                    :  VERTH)/                     %
                    :-----------------------------:  Drilling "PD"
                    :                     (HNTX)   :-----------------------
                    :  Bench work                  %
                    :-----------------------------:  Bench work "PJ"
                    :                         (*)  % :-----------------------
                    :  Control /(KONT OR MTEST)/ %
                    :-----------------------------:  Control "PH"
                    :                   (OVERFL)   :-----------------------
                    :  Surface treatment           %
                    :-----------------------------:  Surface treatment "PG"
                                                    :-----------------------
```

Figure 11.12   The process selection sub-tree.

machine selection. A shaft part, as an example, is shown to indicate this principle in section 11.8.1.

11.7.3   Stock determination

The function of this step is to determine the stock which is suitable for the manufacture of the desired parts. So far, this step is based on the DCLASS raw material database. Here, an algorithm is employed to sort the largest diameter and total length of the workpiece from the part specifications. The value of the largest diameter and total length are then compared with the standard diameters of stock stored in the DCLASS raw material database. Usually, there is a standard diameter of stock which is close to the dimensions of the workpiece; however, two disciplines must be considered in the comparison:

1. There should be enough (but not too much) cutting residue for the maximum diameter of the workpiece.

2. The values of the standard diameters of stock should be consistent with the values issued by the National or International Standard Organization (ISO).

With a large amount of data entered into the database, the structure of the sub-trees are very extended. This situation leads to greater computer memory requirements and larger execution times for traversing the database. In order to overcome this weakness in the current system, a proposal for using R:BASE V to implement the raw material database has been discussed in section 11.6. Two sets of tables, STPD and STRE, have been implemented to represent the final stock specifications (see Figures 11.7 and 11.8). Each specific characteristic of a material defined in a classification sub-tree is represented in each STPD table. According to the specific material characteristic, a code issued from the classification sub-tree is used to search for the corresponding STPD tables. Then two entries, the standard diameters and the material conditions, are used to enter the relevant columns and rows. Here the standard diameter is pre-decided by the traversal of an algorithm sub-tree. This pre-decision is also based on the previous disciplines. The condition of materials is defined in a classification sub-tree, such as black bar, bright bar, hexagon or squared section, etc. Once the two entries are received by the table in STPD, an eight-figure alphanumeric code is issued. The code is used to enter the table in STRE and leads to a row which represents the stock specifications, such as a description of the stock, the dimensions of the stock, the cost per metre, the available situation in storage, and so on. The final stock specification will then be printed out if desired. This table may be considered as a bill of materials interfaced with an MRP system.

11.7.4 Machine selection

Machines are selected automatically for each of the chosen manufacturing processes. Figure 11.13 illustrates a flowchart of this work step. At first the necessary information will be retrieved from the global DCLASS database, such as variables, codes, input keywords, and output keywords. Then a target profile is set up for each process and used to find the candidate machines. The profile consists of three items, a process code, a bit string and some variables. The process code issued from the process selection sub-tree is used to search for machines. The variables are used to avoid machines not capable of producing the part; for example, if a given batch size is required, all machines that are only suitable for smaller batch sizes are ignored. The bit string holds information about the sub-tree traversal and indicates which branches in the process sub-tree were selected. By comparing the bit string from the target profile with the bit string of a machine, the candidate machines

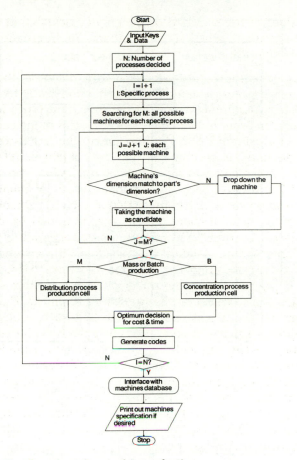

Figure 11.13 The flowchart for machine selection.

can be selected. The candidate machines are then sorted according to a process planning strategy. A strategy may choose a faster machine in preference to a slower machine or a low cost machine in preference to a more expensive one.

The concept of manufacturing cells has been introduced into this step. The machines implemented into the machining database are grouped into several manufacturing cells for different types of production. Currently, ten CNC lathes and seven CNC milling centers have been implemented into the machining database (Figure 11.5). While designing these production cells, two strategies have been followed; the concept of process concentration and process dispersion, and the dimensions of the workpiece. Workpieces are generally divided into three groups: large workpieces, medium workpieces, and small workpieces. As an example, the strategy of grouping ten CNC lathes is illustrated in Figure 11.14.

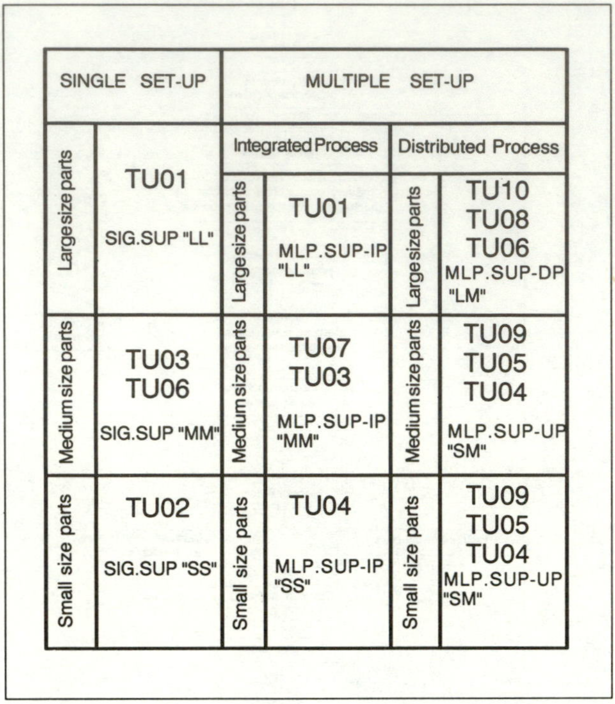

SINGLE SET-UP		MULTIPLE SET-UP			
		Integrated Process		Distributed Process	
Large size parts	TU01 SIG.SUP "LL"	Large size parts	TU01 MLP.SUP-IP "LL"	Large size parts	TU10 TU08 TU06 MLP.SUP-DP "LM"
Medium size parts	TU03 TU06 SIG.SUP "MM"	Medium size parts	TU07 TU03 MLP.SUP-IP "MM"	Medium size parts	TU09 TU05 TU04 MLP.SUP-UP "SM"
Small size parts	TU02 SIG.SUP "SS"	Small size parts	TU04 MLP.SUP-IP "SS"	Small size parts	TU09 TU05 TU04 MLP.SUP-UP "SM"

Figure 11.14 Illustration of strategy of grouping lathes in XPLAN-R.

TUxx are machine ID#s in Figure 11.14, and they can also be seen in Figure 11.5. The codes in each block of Figure 11.14 indicate the grouping situation. The processes for each production cell are alternatives. When the processes are selected, the relevant manufacturing cell is selected. Machines grouped into the cells are flexible and can be adjusted easily to avoid the situation of some machines being heavily tied up by production tasks while others are idle. Here readers should be aware that the functions of concentration and dispersion are different from the concept of alternative process plans. XPLAN-R was developed about six years ago. The integration of process planning with production scheduling was not mature at that time.

11.7.5 Fixture selection

The function of this step is to select suitable fixtures used to clamp the workpiece while processing. The fixtures should also be matched to the machines selected by the system. At the moment, this work step depends on the results carried out from all previous steps, such as the

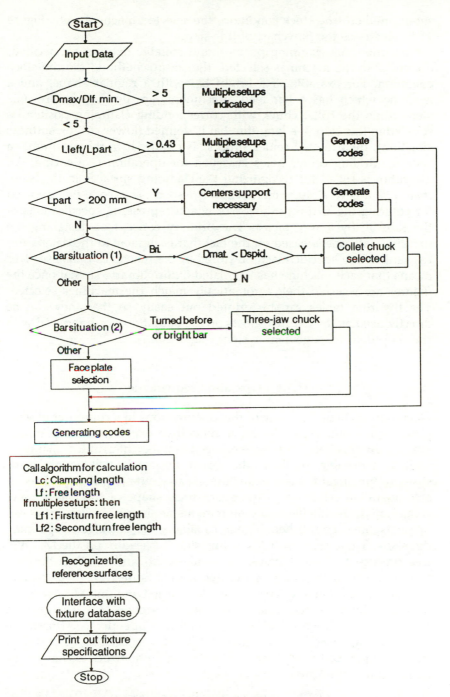

Figure 11.15   The flowchart for fixture selection.

set-up method, the stock condition, the selected machines, etc. Figure 11.15 illustrates the flowchart of this step.

First, the exact clamping position and suitable fixtures are decided, according to the machines selected, the set-up method, and the stock condition. For example, if a bright bar with a regular shape, and a machine which has a bar feed attachment, are decided in previous steps, then the collet chuck with corresponding clamping position is selected in this step. If a long thin bar is defined (lower cutting stiffness situation), then the tail center support is required. In the meantime, a code for drilling center holes is issued for operation sequencing. An algorithm is called for recognizing the clamping surfaces in this step, then a six-figure alphanumeric code is issued. This code will be used for entering the fixturing database, which represents specifications of the R:BASE System V tables. The primary concept of designing the fixturing database is based on the fact that each table in the database is relevant to one specific machine stored in the machines database. This means that each machine has its related fixture library, so that once the fixtures are selected they automatically match the machines selected. The fixturing tables can be printed out easily so that they can be directly used as the fixtures sheets on the shop floor, or interfaced with the shop floor management system.

11.7.6   Operation sequencing

Operation sequencing is one of the key functions in a computer process planning system, especially for a generative expert process planning system. In this case, the system need not be interrupted manually while it is running, unlike in the variant approach. It depends on the system's 'intelligent' knowledge base to carry out an optimal or reasonable operation sequence. If the geometrical shape of the part is more complicated, the intelligence requirements are increased. Recent developments have established higher capability for intelligent operating decisions. Thus far, in XPLAN-R this step is based upon the DCLASS tree processor. A combination of a rule-based and an algorithm approach are used to sequence the operations. Some very detailed and specified operation sequencing can be carried out by XPLAN-R, as illustrated by the process plans presented in section 11.8. The detailed flowchart for this work step is illustrated in Appendix 2. At first, the rule-based approach, which makes use of a collection of IF-THEN rules, is employed to register how many different attributes have to be operated upon, for instance:

IF: external   THEN: sorting external diameters

IF: internal   THEN: sorting internal diameters

IF: grooves   THEN: sorting grooves

IF: chamfers   THEN: sorting chamfers

IF: . . .   THEN: . . .

But these are not sufficient to generate a detailed operation sequence. The variation of external or internal diameters can be in any order in mechanical designing; for example, a variety of diameters can be obtained by decreasing or increasing from one side, or from both sides, or by using some other methods. A few smaller algorithm sub-trees are employed here to classify the different situations; for example, the order of external or internal diameters is sorted and rearranged by the algorithms. The rearrangements depend upon the concept of optimal manufacturing technology, then the IF-THEN logic processing capabilities are used again to effectively output a series of sequenced CODEs which define the operation sequence. These CODEs can be used in conjunction with the database and a report generator; here it is the XPLANR.INP file (Zhang, XPLAN-R Reference Manual, 1989). First the associated operation data, such as dimensions, cutting reference coordinates, and so on, are picked up from the configuration database, then the corresponding cutting tools and fixtures are picked up from the tools and fixtures database, and finally an operation plan sheet is generated.

11.7.7   Tooling selection

The function of this step is to select cutting tools which are appropriate to each operation sequence and cutting condition. At the moment, this work step is independently based upon the DCLASS tree processor. For each specific operation, a code is issued; this code will then lead to a variable which, at the end of the work step, yields a candidate cutting tool. For instance, if, in the operation sequencing step, a sidebit turning operation is defined for an external diameter, the code 'sidebit' is issued. This code leads to a variable, e.g. EXTO (I). Then, at the end of the step, the variable EXTO (I) points to an external sidebit cutter. In the same procedure, an insert turn operation will generate an insert cutter, and an internal finished turn will generate an internal finishing cutter. However, some weakness exists in this step. So far, the work step only takes into account the factors of the operation sequencing and some cutting conditions, but not the factor of cutting materials, which is usually an important factor for the selection of cutting tools. For example, for an identical external sidebit turn, different part materials such as high carbon steel, bronze, or gray cast iron may have much different requirements for the cutting tool selection.

A proposal for improving the capacity and capability of this step has

Most often used raw materials

	Alloy steel	Aluminium alloy	Carbon steel	Composite/ Plastic	Copper/ Bronze	Gray cast iron	Stainless steel	. . . .
CF92								
FL11								
FS12								
FT52								
RS11			TC90CS			TC90GC		
. . .								
XX00	XX00XX	XX00XX	XX00XX	XX00XX	XX00XX	XX00XX	XX00XX	. . . . .

CODES I X X 0 0: Four alphanumeric code to indicate the cutting subjects
     →Codes of cutting dimensions group
     →Codes of cutting surface shape
     →Codes of process activities

CODES	Tools specifications	Conditions			Cost/ min.	Other necessary information	Codes for tools manage-ment	Tools situa-tion
		max. feed	max. depth	max. speed				
XX00XX								
TC90CS	Tungsten sidebit turn for carbon steel with 90 angle	0.3/S	5mm	60m/ min	20DDK/ min		TC90	Y
TC90GC	Tungsten sidebit turn for gray cast iron with 90 angle	1.5/S	8mm	40m/ min	12DDK/ min		TC90	Y
. . .								
XX00XX								

CODES II XX00XX: Six-alphanumeric-code to indicate the tool conditions
     →Codes of workpiece materials
     →Codes of tool main geometry
     →Codes of tool materials

Figure 11.16 An example of tooling selection tables.

been established. The primary concept of designing this step adopts the same idea as stock determination. An R:BASE V tooling database is assumed to be interfaced with XPLAN-R, which is accessed as follows. The relevant R:BASE V tooling databases consist of two kinds of tables, namely COGE (COdes GEneration) tables and TOSP (TOols Specification Presentation) tables. Each machine corresponds to its own COGE table. This means that each machine has its related tooling library. An example of such tables is shown in Figure 11.16.

The function of COGE tables is to preselect tools by means of issuing codes. The first column of the table is a group of four-figure alphanumeric codes defined here as CODE (I). The CODE (I) issued from the operation sequencing indicates the cutting subjects. For example, the code in the fifth row in Figure 11.16 is RS11. The two letters are the codes of process activities; specifically here, RS means Rough, Sidebit turn. The first digit is the code for cutting surface shape; here number 1 means an external diameter surface. The last digit is the code for the cutting dimensions of the diameter group. The range of diameters is divided into several groups; specifically here, number 1 means that the range of cutting diameters is from 5 mm to 30 mm. The detailed description of these codes can be found in Appendix C. The first row of the table COGE is used to store some often used raw materials such as alloy steel, carbon steel, gray cast iron, etc., specified in the work step of the part specification. Once the above elements, the specific material, and the code of the cutting subject are received, the six-figure alpha-numeric code defined as CODE (II) is issued. For example, if CODE (I) RE11 and the raw material carbon steel are received, then the CODE (II) TC90CS is issued from the table COGE. The CODE (II) TC90CS is then used for entering the table TOSP. The function of the table TOSP is to represent the specifications of tools selected by the system. CODE (II) leads to a row which provides the detailed specifications of the tools. For instance, CODE (II) TC90CS means that this is an external sidebit cutter with a 90 degree cutting angle, the material of the tool is tungsten alloyed tool steel, and the cutting material is medium carbon steel. The cutting condition is also provided in this row (see Figure 11.16). A column entitled 'Codes for tools management' is included in the table TOSP and is especially useful for tool management. One can see that, in the TOSP table in Figure 11.16, rows 3 and 4 are for identical process activities but for two different cutting materials (one for carbon steel, the other for gray cast iron), for which table TOSP provides an identical cutting tool but different cutting conditions. The table TOSP can be considered as a tooling sheet used on a shop floor or interfaced with a tooling management system.

11.7.8 Cutting condition decision

Cutting condition decision is associated with the tooling selection. The specification data for tooling will be retrieved in this step, then an algorithm is called to calculate the cutting conditions. So far, the concept for the calculation of the tolerance of removal operations is proposed in this step, but the real subroutines have not yet been implemented. The results of this step are assumed to be easily transformed into an NC path. Figure 11.17 illustrates the flowchart for this step.

11.7.9 Time and cost calculation

Each company has its own rules for calculating the machining time consumption; the time calculation sub-tree has been built up very

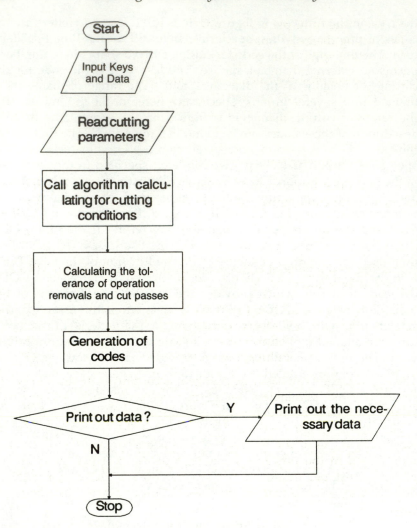

Figure 11.17   The flowchart for cutting condition decisions.

flexibly in XPLAN-R. A more accurate time calculation algorithm has been implemented into the system in the form of a DCLASS tree structure. This calculation is based on the function

$$T = \frac{L}{N \cdot S} = \frac{\pi \cdot D \cdot L}{V \cdot S} \qquad \text{(F.11-1)}$$

where:
   T: Machining time for each cut (minutes)
   L: Cutting length (mm)
   N: Spindle speed (rpm)

S: Feed rate (mm per revolution)
D: Cutting diameter (mm)
V: Cutting speed (mm per minute)

In order to simplify the calculation, a time parameter P is defined as

$$P = \frac{V \cdot S}{\pi} \tag{F.11-2}$$

where:
  P: Time parameter (mm^2 per minute)

If we introduce function (F.11-2) into function (F.11-1), the final time calculation function employed in XPLAN-R is derived in the form

$$T = \frac{D \cdot L}{P} \tag{F.11-3}$$

According to the general cutting condition, the value of P in XPLAN-R is defined as follows:

$$P \text{ rough, sidebit turn} = 3000 \tag{F.11-4}$$
$$P \text{ rough, insert turn} = 2500 \tag{F.11-5}$$
$$P \text{ finish turn} = 1500 \tag{F.11-6}$$
$$P \text{ threading} = 1000 \tag{F.11-7}$$

These parameters can easily be changed, according to the user's specific strategy. Time parameters for cutting features, such as chamfers, knurls, etc., have been defined in each specific sub-tree.

The algorithm for cost calculation has not yet been implemented. Nevertheless, all of the necessary fundamental parameters have been represented in the previous steps, such as the cost per meter of raw material, cost per minute of tool wear, cost per minute of machine depreciation, etc., so that the final calculation of overall cost should be easy.

11.8   TEST EXAMPLES

The architecture of XPLAN-R and the detailed working steps have been introduced in the previous part of this chapter. Now, in order to demonstrate the capacity and capability and the knowledge of manu-facturing logic and some of the production know-how for rotational components from real production environments, which have been implemented into the system, three examples are tested here. The criteria for selecting these three examples is that they should be:

1. as close to the actual production environments as possible
2. configured in complex geometrical shapes

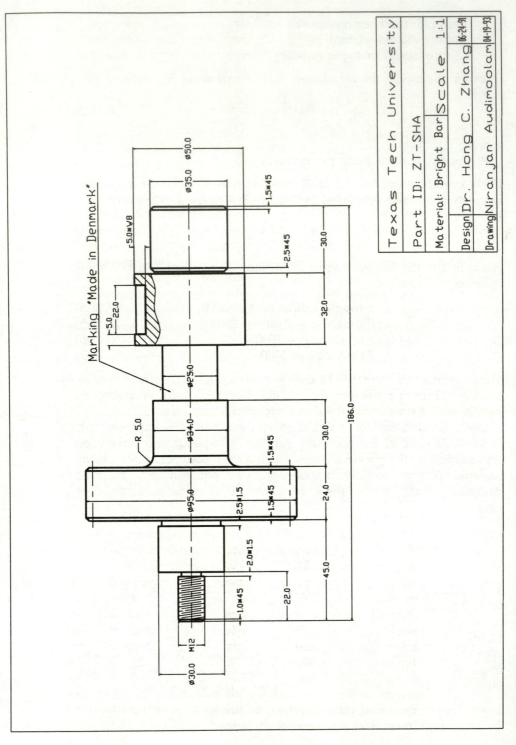

Figure 11.18 An example of a shaft part (ZT-SHA).

Testing these three examples will not only demonstrate the primary principles of XPLAN-R, but will also illustrate the basis of the strategies for the recognition of the workpiece by means of the fuzzy boundary part family.

11.8.1 An example of a shaft part

The first example belongs to the shaft part group illustrated in Figure 11.18. The ID# of the part is ZT-SHA. It is a gear-shaft with seven diameters, all arranged irregularly, numbered from left to right. The largest diameter is No. 3 and the smallest diameter is No. 1, which is threaded. Two diameters, diameters No. 4 and No. 5, are smaller than the adjacent diameters. There is a keyway on diameter No. 6. A test marking, 'MADE IN DENMARK', is required in a vertical position on the axis of the part. Several grooves, fillets, and chamfers are required on different diameters, as illustrated in Figure 11.18. This example is to demonstrate how XPLAN-R processes shaft parts and also to show how it processes features. With two different types of production batch sizes, the concepts of process concentration and process dispersion are also illustrated in two respective plan sheets. The geometrical components of the example are input from the left end to the right end of the part in terms of diameter numbers (from No. 1 to No. 7). Figure 11.19 is the screen display of the first few steps of the interactive dialogue between the system and the user. Note that the production batch size of 100 is regarded as a quantity production type in XPLAN-R. The complete screen display of the interactive input for the part ZT-SHA is presented in Appendix D. The corresponding process plan sheet is shown in Figure 11.20.

In Figure 11.20, P# stands for process numbers and O# stands for operation numbers. Because of the complex shape of the part ZT-SHA, the 'centers with tailstock supported' clamping method was selected, so that the rough turning was considered as two processes: first rough turn and second rough turn. A total of six processes were selected automatically by XPLAN-R; they are P#10 raw material preparation, P#20 rough turn right part of ZT-SHA (first rough turn), P#30 rough turn left part of ZT-SHA (second rough turn), P#40 finish turn with double centers supporting, P#50 milling keyway, and P#60 final inspection. In the raw materials storage stage, bright bar with diameter D = 98 mm and length L = 191 mm, which are optimum material data (with minimum cutting residual), were selected. Three different turning machines, namely YAMAZAKI QUICK TURN 20, OKUMA LB15-1SC 500, and MORI SEIKI CNC SL-5, were selected for three different turning processes. The selection of machines is based on the logic of manufacturing cells. The method used for one-sided clamping is chuck clamping, and the other side is center supported for the two

```
                           X P L A N - R
                      Process Planning System
                           version 2.0

                         (C) Copyright by
                   Institute for Product Development
               Laboratory of Process and Production Engineering
               Building 425, Technical University of Denmark

        Enter name
      >>H.C. Zhang

        Enter Password

      >> XPLAN-R Main menu
      1. Generative Process Planning
      2. Print last generated Plan
      3. Plan with alternative machines
      4. Options
      5. Database management
      99. EXIT
      >>1

        Enter ID-number >>ZT-SHA

        Enter Description : Shaft group parts
        >>

        Main shape
    "   1 - Rotational
        2 - Prismatic
        3 - Other mainshape
    "",
          - Enter batch size
                 Value = 100
      >>100

        Material
    "   1 - Metal
        2 - Plast
        3 - Other
    "",

        Metal
    "   1 - Steel
        2 - Aluminum
        3 - Brass MS58
        4 - Inconel
        5 - Berylco
        6 - Other
    "",

        Steel
    "   1 - Automate steel
        2 - Stainless 43
    "",

        Condition
    "   1 - Bar
        2 - Semi product
        3 - Other
    "".
```

Figure 11.19 The screen display of the first input steps.

rough turn processes. Besides the necessary process data, such as machines, clamping tools, etc., the necessary operation data are also decided by the system; for example, in Figure 11.20, P#20 and O#60 are for rough turning diameter No. 6. The turning diameter is 50 mm and the turning length is 117 mm. The turning is started from the right side of the part, where the initial point of the reference operation

```
-------------------------------------------------------------------
                         X P L A N - R              1989- 7-31
                         PROCESS  PLAN
   Part ID:ZT-SHA                      Planner:H.C. Zhang
-------------------------------------------------------------------
   P#  O#   Description              Tools        Setup T. Mach.T.


   10 - - - - - - - - - - - - - - - - - - - - - - - - - - - - - -
         10  Raw material storage
             Automate st bar D=98.00   L=191.00
   20 - - - - - - - - - - - - - - - - - - - - - - - - - - - - - -
             YAMAZAKI Quick T.20                    .45   h  .14  h
         10  Clamp part with ext.L=153.50  Chuck
         20  Facing section  depth=2.50    Facer
         30  Drill center    depth=8.50mm  Cent. Driller
         40  Tailstock & Center surpoting  Center
         50  Rough turn diameter D=95.00   Sidebit         4.46  m
                                 X=0.00      L=141.00
         60  Rough turn diameter D=50.00   Sidebit         1.95  m
                                 X=0.00      L=117.00
         70  Rough turn diameter D=35.00   Sidebit          .35  m
                                 X=0.00      L=30.00
         80  Rough turn diameter D=34.00   Insertin         .34  m
                                 X=87.00     L=30.00
         90  Rough turn diameter D=25.00   Insertin         .20  m
                                 X=62.00     L=25.00
        100  Turn square groove  B=2.50    Grosqubit        .56  m
                                 X=32.50     D=1.50
        110  Turn right chamfer  A=45.00   Sidebit          .26  m
                                 X=117.00    W=1.50
        120  Turn right chamfer  A=45.00   Sidebit          .15  m
                                 X=0.00      W=1.50
        130  Turn left fillet    R=5.00    Fit.bit          .48  m
                                 X=117.00
   30 - - - - - - - - - - - - - - - - - - - - - - - - - - - - - -
             OKUMA LB15-1SC 500                     .35   h  .03  h
         10  Clamp part with ext.L=119.50  Chuck
         20  Facing section  depth=2.50    Facer
         30  Drill center    depth=8.50mm  Cent. Driller
         40  Tailstock & Center surpoting  Center
         50  Rough turn diameter D=30.00   Sidebit          .45  m
                                 X=0.00      L=45.00
         60  Rough turn diameter D=12.00   Sidebit          .08  m
                                 X=0.00      L=22.00
         70  Turn square groove  B=2.00    Grosqubit        .54  m
                                 X=20.00     D=1.50
         80  Turn square groove  B=2.50    Grosqubit        .58  m
                                 X=42.50     D=1.50
         90  Turn left  chamfer  A=45.00   Sidebit          .15  m
                                 X=0.00      W=1.00
        100  Turn left  chamfer  A=45.00   Sidebit          .19  m
                                 X=45.00     W=1.50
   40 - - - - - - - - - - - - - - - - - - - - - - - - - - - - - -
             MORI SEIKI CNC SL-5                    .50   h  .26  h
         10  Double centers surpoting with Cen. & Tai.
             length of part L=186.00
         20  Fine turn diameter  D=12.00   Finebit          .17  m
                                 X=0.00      L=22.00
         30  Fine turn diameter  D=30.00   Finebit          .46  m
                                 X=22.00     L=23.00
         40  Fine turn diameter  D=95.00   Finebit         1.52  m
                                 X=45.00     L=24.00
         50  Fine turn diameter  D=34.00   Finebit          .68  m
                                 X=69.00     L=30.00
         60  Fine turn diameter  D=25.00   Finebit          .41  m
                                 X=99.00     L=25.00
         70  Fine turn diameter  D=50.00   Finebit         1.06  m
                                 X=124.00    L=32.00
         80  Fine turn diameter  D=35.00   Finebit          .70  m
                                 X=156.00    L=30.00
         90  Turn ext. thread    D=12.00   Threadbit      10.00  m
                                 X=166.00    L=12.00
        100  Turn marking: MADE IN DENMARK Text wheels     1.00  m
   50 - - - - - - - - - - - - - - - - - - - - - - - - - - - - - -
             CORTINI L300 Center                    .60   h  .04  h
         10  Mill keyway         B=8.00    Endmill A       2.40  m
                                 D=5.00      L=22.00

-------------------------------------------------------------------
```

Figure 11.20 Plan sheet for process dispersion of (ZT-SHA).

coordinate is placed. This means that the turning tool for diameter No.
6 is started from the zero point and the moving length is 117 mm. The
corresponding cutting tool is a 'sidebit' with cutting time 1.95 minutes
calculated. In the same process number, operation O#90 is for rough

```
Enter ID-number  >>ZT-SHA2

Enter Description :
>>Shaft group parts for batch size production

Main shape
1 - Rotational
2 - Prismatic
3 - Other mainshape
**>1
- Enter batch size
>>10

Material
1 - Metal
2 - Plast
3 - Other
**>1

Metal
1 - Steel
2 - Aluminum
3 - Brass MS58
4 - Inconel
5 - Berylco
6 - Other
**>
```

Figure 11.21   Interactive input for batch size production.

turning diameter No. 5. The turning diameter is 25 mm and the turning length is 25 mm. The value of the operation coordinate is 62 mm. This means that the turning tool is started from a position 62 mm from the initial point of the reference coordinate. The corresponding tool is 'insert' with cutting time 0.56 minute. The optimized manufacturing decision rules have indicated that the all sidebit turning should take place before any insert turning and that sidebit turning should remove as much as possible. Therefore changing tools from sidebit to insert occurs only once, and the consecutive insert turning depths for the front of the diameters will be reduced. This means that the results of the process planning coincide with the optimized manufacturing decision logic.

It is obvious that for the quantity production size the processes are distributed. However, for the same part, when the production size requires batch size production (batch number 10 is indicated in Figure 11.21), then the corresponding process plan sheet is as shown in Figure 11.22.

In Figure 11.22, a single lathe, BOEHRINGER VDF 400C, is employed to replace all three lathes which carry out the turning processes in Figure 11.20. This indicates that XPLAN-R can distinguish automatically either process concentration or dispersion based on the types of production.

```
-------------------------------------------------------------------------
                            X P L A N - R              1989- 7-31
                            PROCESS  PLAN
        Part ID:ZT-SHA                     Planner:H.C. Zhang
        -----------------------------------------------------------------
         P#  O#   Description                   Tools        Setup T. Mach.T.
        -----------------------------------------------------------------

         10 - - - - - - - - - - - - - - - - - - - - - - - - - - - - - - -
                 Raw material storage
             10  Automate st bar D=98.00    L=191.00
         20 - - - - - - - - - - - - - - - - - - - - - - - - - - - - - - -
                 NAKAMURA Super T.2B                          .80   h  .44  h
             10  Clamp part with ext.L=153.50    Chuck
             20  Facing section   depth=2.50     Facer
             30  Drill center     depth=8.50mm   Cent. Driller
             40  Tailstock & Center surpoting    Center
             50  Rough turn diameter D=95.00     Sidebit          4.46  m
                     X=0.00          L=141.00
             60  Rough turn diameter D=50.00     Sidebit          1.95  m
                     X=0.00          L=117.00
             70  Rough turn diameter D=35.00     Sidebit           .35  m
                     X=0.00          L=30.00
             80  Rough turn diameter D=34.00     Insertin          .34  m
                     X=87.00         L=30.00
             90  Rough turn diameter D=25.00     Insertin          .20  m
                     X=62.00         L=25.00
            100  Turn square groove  B=2.50      Grosqubit         .56  m
                     X=32.50         D=1.50
            110  Turn right chamfer  A=45.00     Sidebit           .26  m
                     X=117.00        W=1.50
            120  Turn right chamfer  A=45.00     Sidebit           .15  m
                     X=0.00          W=1.50
            130  Turn left fillet    R=5.00      Fit.bit           .48  m
                     X=117.00
            140  Rotate and reclamp the part     Chuck
                 with extend        L=119.50
            150  Facing section   depth=2.50     Facer
            160  Drill center     depth=8.50mm   Cent. Driller
            170  Tailstock & Center surpoting    Center
            180  Rough turn diameter D=30.00     Sidebit           .45  m
                     X=0.00          L=45.00
            190  Rough turn diameter D=12.00     Sidebit           .08  m
                     X=0.00          L=22.00
            200  Turn square groove  B=2.00      Grosqubit         .54  m
                     X=20.00         D=1.50
            210  Turn square groove  B=2.50      Grosqubit         .58  m
                     X=42.50         D=1.50
            220  Turn left  chamfer  A=45.00     Sidebit           .15  m
                     X=0.00          W=1.00
            230  Turn left  chamfer  A=45.00     Sidebit           .19  m
                     X=45.00         W=1.50
            240  Reclamp the part with double    Cen. & Tai.
                 centers surpoting  L=186.00
            250  Fine turn diameter  D=12.00     Finebit           .17  m
                     X=0.00          L=22.00
            260  Fine turn diameter  D=30.00     Finebit           .46  m
                     X=22.00         L=23.00
            270  Fine turn diameter  D=95.00     Finebit          1.52  m
                     X=45.00         L=24.00
            280  Fine turn diameter  D=34.00     Finebit           .68  m
                     X=69.00         L=30.00
            290  Fine turn diameter  D=25.00     Finebit           .41  m
                     X=99.00         L=25.00
            300  Fine turn diameter  D=50.00     Finebit          1.06  m
                     X=124.00        L=32.00
            310  Fine turn diameter  D=35.00     Finebit           .70  m
                     X=156.00        L=30.00
            320  Turn ext. thread    D=12.00     Threadbit       10.00  m
                     X=166.00        L=12.00
            330  Turn marking: MADE IN DENMARK   Text wheels      1.00  m
         30 - - - - - - - - - - - - - - - - - - - - - - - - - - - - - - -
                 CORTINI L300 Center                          .60   h  .04  h
             10  Mill keyway         B=8.00      Endmill A              2.40  m
                     D=5.00          L=22.00

        -----------------------------------------------------------------
-------------------------------------------------------------------------
```

Figure 11.22   Plan sheet for process concentration of (ZT-SHA).

11.8.2   An example of a screw part

The second example belongs to a screw part group and is illustrated in Figure 11.23. The ID number is ZT-SCR. There are five diameters numbered from the left end to the right end; diameter No. 2 is a concave curve, diameter No. 4 is a straight taper, the others are cylindrical

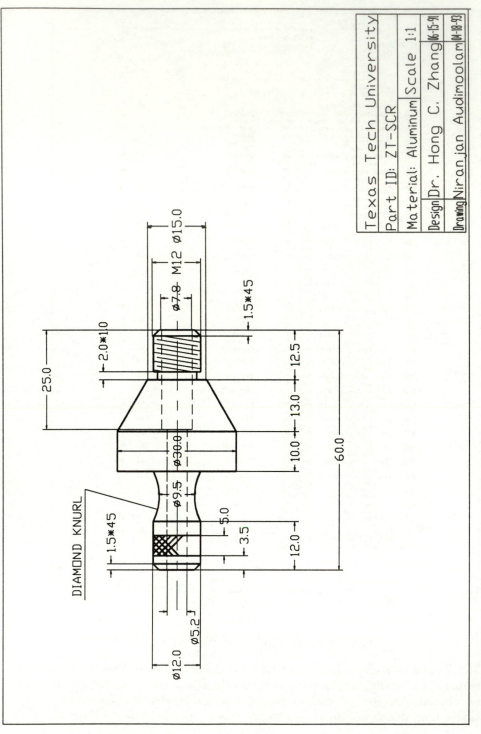

Figure 11.23 An example of a screw part (ZT-SCR).

```
-------------------------------------------------------------------
                         X P L A N - R           1989- 7-31
                         PROCESS PLAN
        Part ID:ZT-SCR                        Planner:H. C. Zhang
        -----------------------------------------------------------
         P#  O#   Description                Tools        Setup T. Mach.T.
        -----------------------------------------------------------

         10 - - - - - - - - - - - - - - - - - - - - - - - - - - -
                  Raw material storage
             10   Aluminum   bar D=32.00    L=98.47
         20 - - - - - - - - - - - - - - - - - - - - - - - - - - -
                  COLCHESTER CNC 2000L                   .40   h  .16  h
             10   Clamp part with ext.L=82.50   Chuck
             20   Facing section  depth=2.50    Facer
             30   Rough turn diameter D=30.00   Sidebit        .60   m
                  X=0.00          L=60.00
             40   Rough turn diameter D=30.00   Sidebit        .25   m
                  X=0.00          L=25.50
             50   Rough turn diameter D=12.00   Sidebit        .05   m
                  X=0.00          L=12.50
             60   Rough turn diameter D=12.00   Insertin       .06   m
                  X=35.50         L=12.50
             70   Rough turn diameter D=12.00   Insertin       .05   m
                  X=48.00         L=12.00
             80   R. turn posi. taper Dmax=30.00  Sidebit
                  X=12.50  L=13.00  Dmin=15.00  Tag.=1.15
             90   R. T. convave curd. Dmax=12.00  Cocav.bit
                  X=35.50  L=12.50  Dmin=9.50
            100   R. dri. inte. diam. D=5.20    Driller        1.20  m
                  X=0.00          L=60.00
            110   R. bor. inte. diam. D=7.80    Borebit
                  X=0.00          L=25.00
            120   Turn square groove  B=2.00    Grosqubit      .52   m
                  X=14.50         D=1.00
            130   Fine turn diameter  D=12.00   Finebit        .10   m
                  X=0.00          L=12.50
            140   F. turn posi. taper Dmax=30.00  Finebit
                  X=12.50  L=13.00  Dmin=15.00  Tag.=1.15
            150   Fine turn diameter  D=30.00   Finebit        .20   m
                  X=25.50         L=10.00
            160   F. T. convave curd. Dmax=12.00  Cocav.bit
                  X=35.50  L=12.50  Dmin=9.50
            170   Fine turn diameter  D=12.00   Finebit        .09   m
                  X=48.00         L=12.00
            180   F. turn ins. diam.  D=7.80    Finebit        1.00  m
                  X=0.00          L=25.00
            190   F. turn ins. diam.  D=5.20    Finebit
                  X=25.00         L=35.00
            200   Turn diamond knurl  P=1.00    d.Knl.wheels   .20   m
                  X=51.50         W=5.00
            210   Turn left  chamfer  A=45.00   Sidebit        .21   m
                  X=60.00         W=1.50
            220   Turn right chamfer  A=45.00   Sidebit        .15   m
                  X=0.00          W=1.50
            230   Turn ext. thread    D=12.00   Threadbit      5.00  m
                  X=.50           L=12.00
            240   Cut off the part    L=60.00   Cutter         .40   m
                  X=0.00
         30 - - - - - - - - - - - - - - - - - - - - - - - - - - -
                  Control Department
             10   Inspection

        -----------------------------------------------------------
```

Figure 11.24 The process plan sheet for (ZT-SCR).

shapes. Some other features are shown on the part too; for example, a left chamfer and a diamond knurl are on diameter No. 1, a thread, a square groove and a right chamfer are on diameter No. 5. The internal diameters are stepped. This example demonstrates how XPLAN-R processes the curved or tapered external features. The corresponding process plan is shown in Figure 11.24.

The plan includes the machines, the selected processes, the operation sequences, and the necessary clamping approach, clamping tools, and clamping data. The corresponding cutting data for each operation are all defined, such as cutting dimensions (diameter and length), the values of the cutting reference coordinate, as well as cutting tools. The approximate machining time is also calculated by XPLAN-R.

A mechanism for alternative machine selection has been built in XPLAN-R so that the machines selected by the system can be replaced by alternatives. This function can be used to avoid the situation that some machines are fully loaded and some are underused. It can also be used for adjustment when some machines are temporarily being repaired or not available within a production cell. The alternative machine selection can very easily be made by selecting option No. 3 in the XPLAN-R main menu (Zhang, XPLAN-R Reference Manual, 1989). An example of this function and the corresponding plan sheet are illustrated in Appendix D, where a MORI SEIKI CNC SL-5 lathe is employed to replace the COLCHESTER CNC 2000L lathe shown in Figure 11.24.

11.8.3   An example of a flange part

The third example belongs to a flange part group and is illustrated in Figure 11.25. The ID number is ZT-FLA. There are five external diameters and three internal diameters, with six non-coaxial holes symmetrically located on the disc of the flange as well as a hole with diameter 6 mm which forms a 45 degree angle with the part's axis. This example demonstrates how XPLAN-R processes internal diameters and features. The corresponding process plan is shown in Figure 11.26. The detailed screen displays and layouts for all of the above examples are shown in Appendix D.

11.9   CONCLUSION

XPLAN-R, as a prototype of an expert process planning system, has been discussed in detail. Three examples have been tested and the corresponding process plans are obviously close to real production environments. However, as mentioned previously, XPLAN-R was developed about six years ago, when the development technology for CAPP was not very mature and there were limitations in terms of hardware and software.

We have reviewed the advantages of XPLAN-R, but readers should also be able to point out the obvious drawbacks of XPLAN-R, based on the discussion in previous chapters. The greatest drawback of XPLAN-R is that the system does not integrate with CAD or production

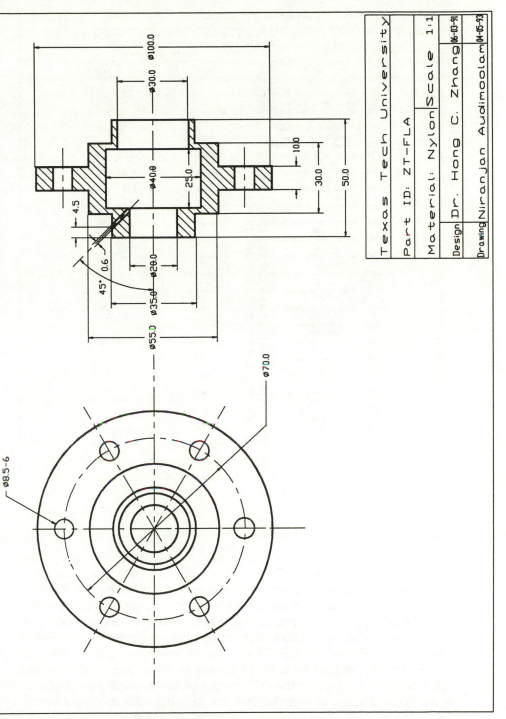

Figure 11.25 An example of a flange part (ZT-FLA).

```
---------------------------------------------------------------
                         X P L A N - R                1989- 7-31
                         PROCESS PLAN
      Part ID:ZT-FLA                     Planner:H.C. Zhang
      ---------------------------------------------------------
      P#  O#   Description                Tools        Setup T. Mach.T.
      ---------------------------------------------------------

      10 - - - - - - - - - - - - - - - - - - - - - - - - - - - -
             Raw material storage
          10   Nylon        bar D=100.00   L=85.85
      20 - - - - - - - - - - - - - - - - - - - - - - - - - - - -
             COLCHESTER CNC 2000L                        .40  h  .13  h
          10   Clamp part with ext.L=72.50   Chuck
          20   Facing section  depth=2.50    Facer
          30   Rough turn diameter D=100.00  Sidebit          1.66  m
               X=0.00          L=50.00
          40   Rough turn diameter D=55.00   Sidebit           .36  m
               X=0.00          L=20.00
          50   Rough turn diameter D=35.00   Sidebit           .11  m
               X=0.00          L=10.00
          60   Rough turn diameter D=55.00   Insertin          .22  m
               X=30.00         L=10.00
          70   Rough turn diameter D=35.00   Insertin          .14  m
               X=40.00         L=10.00
          80   R. dri. inte. diam. D=20.00   Driller          1.80  m
               X=0.00          L=50.00
          90   R. bor. inte. diam. D=30.00   Borebit
               X=0.00          L=35.00
         100   R. bor. inte. diam. D=40.00   Insertin
               X=15.00         L=20.00
         110   Fine turn diameter D=35.00    Finebit           .23  m
               X=0.00          L=10.00
         120   Fine turn diameter D=55.00    Finebit           .36  m
               X=10.00         L=10.00
         130   Fine turn diameter D=100.00   Finebit           .66  m
               X=20.00         L=10.00
         140   Fine turn diameter D=55.00    Finebit           .36  m
               X=30.00         L=10.00
         150   Fine turn diameter D=35.00    Finebit           .23  m
               X=40.00         L=10.00
         160   F. turn ins. diam.  D=30.00   Finebit          1.50  m
               X=0.00          L=15.00
         170   F. turn ins. diam.  D=40.00   Finebit
               X=15.00         L=20.00
         180   F. turn ins. diam.  D=20.00   Finebit
               X=35.00         L=15.00
         190   Cut off the part   L=50.00    Cutter            .40  m
                                X=0.00
      30 - - - - - - - - - - - - - - - - - - - - - - - - - - - -
             HERMLE UWF 700 CNC                          .45  h  .14  h
          10   Drill 6      holes  d=8.5     Drill d=8.5  mm   7.65  m
                                X=38.5
          20   Drill 1      holes  d=6       Drill d=6    mm   1.20  m
                                A=45
      ---------------------------------------------------------------
```

Figure 11.26   The process plan sheet for (ZT-FLA).

scheduling, since the design information is input in interactive form. Although the system can generate alternative process plans by means of different production types, the system can produce neither non-linear process plans nor dynamic process plans in terms of integrability. XPLAN-R is based on the DCLASS tree processor and a relational database R:BASE System V. DCLASS was so popularly used at the end of the 1980s because of the advantages of the explicit tree structure. According to recent reports, the use of DCLASS has significantly declined because of the shortage of object-oriented systems. DCLASS is implemented in FORTRAN 77, which is definitely an obsolete programming language for an integrated system; C/C++ predominate for the implementation of such systems, but there is no evidence that DCLASS will be rewritten in C/C++ soon. This is an obvious drawback for integrating with other object-oriented systems. The relational database R:BASE V has the same problems. In Chapter 10, we stated that the development trends of databases for CAPP are object-oriented databases that are easy to integrate into other systems, but R:BASE V does not prossess this function. So in terms of software, XPLAN-R again falls short. In addition, although some idea of tolerance analysis has been introduced into the system, the analytical function falls short of the needs of users; in particular, automated tolerance analysis has not been implemented in the system.

In this chapter we have discussed an expert process planning system XPLAN-R. The advantages and disadvantages have been discussed in detail. Readers may use these discussions to develop their own system or to adapt an existing system. As the last sentence of this book, we hope that you can carry out a splendid computer aided process planning system based on our contribution.

REFERENCES

Alting, L., Zhang, H.C. and Lenau, T. (1988) XPLAN – An Expert Process Planning System and its Further Development. Proceedings of 27th International MATADOR Conference, UMIST, UK, 20–21 April.

Alting, L. and Zhang, H.C. (1989) Computer aided process planning: the state-of-the-art survey. *International Journal of Production Research*, **27**(4).

DCLASS Technical Manual (1985) CAM Software Research Laboratory, Brigham Young University, Provo, Utah.

Lenau, T. and Alting, L. (1986) XPLAN – An Expert Process Planning System. Second International Expert Systems Conference, London, 30 September–2 October.

R:BASE System V (1986) *Learning Guide*, Microrim Inc.

Zhang, H.C. (1987) *Overview of CAPP and Development of XPLAN*. Institute of Manufacturing Engineering, Technical University of Denmark, Publication No. AP.87-26.

Zhang, H.C. (1989) *XPLAN-R Reference Manual*, Institute for Production Development, Laboratory of Process and Production Engineering, Technical University of Denmark, Lyngby, Publication No. PI.89.18-A.

Zhang, H.C. and Alting, L. (1988a) XPLAN-R – An Expert Process Planning System for Rotational Components. Proceedings of IIE 1988 Integrated Systems Conference, St Louis, Missouri, USA, October 30–November 2.

Zhang, H.C. and Alting, L. (1988b) Introduction to an Intelligent Process Planning System for Rotational Parts. ASME-WAM 1988 PED-vol. 31, Chicago, Illinois, November 27 – December 2.

Zhang, H.C. and Alting, L. (1989) Expert Process Planning Systems: The State-of-the-art. Proceedings of 21st CIRP International Seminar on Manufacturing Systems, Stockholm, Sweden, June 5–6.

List of raw materials in the database

S T O C K   I N V O I C E   DATE: 01/24/89

* * * * * * * * * * * * *

Institute of Manufacturing Engineering
Technical University of Denmark
DK-2900 Lyngby, Denmark

This STOCK INVOICE is used for the RWA MATERIALS DATABASE
of the expert process planning system XPLAN-R

Property of stock: Aluminium 13

Codes	Features	Dia. (mm)	Length (m)	Cost/m (DKK)	Stock state	Location
BRL13030	Bright bar	30.	8.	140.	Available	AL.02.6
BRL13035	Bright bar	35.	8.	150.	Available	AL.02.7
BRL13040	Bright bar	40.	8.	150.	Available	AL.02.8
BRL13045	Bright bar	45.	8.	155.	Available	AL.02.9
HXL13020	Hexagonal bar	20.	5.	180.	Available	AL.03.4
HXL13025	Hexagonal bar	25.	5.	185.	-0-	AL.03.5
HXL13030	Hexagonal bar	30.	5.	245.	Available	AL.03.6
HXL13035	Hexagonal bar	35.	5.	250.	-0-	AL.03.7
HXL13040	Hexagonal bar	40.	5.	265.	Available	AL.03.8
HXL13045	Hexagonal bar	45.	-0-	-0-	-0-	-0-
SQL13020	Squaredsection	20.	5.	230.	Available	AL.04.5
SQL13025	Squaredsection	25.	-0-	-0-	-0-	-0-
SQL13030	Squaredsection	30.	5.	280.	Available	AL.04.5
SQL13035	Squaredsection	35.	-0-	-0-	-0-	-0-
SQL13040	Squaredsection	40.	-0-	-0-	-0-	-0-
SQL13045	Squaredsection	45.	6.	-0-	-0-	-0-
BLL13020	Black bar	20.	8.	70.	Available	AL.01.4
BLL13025	Black bar	25.	8.	75.	Available	AL.01.5
BLL13030	Black bar	30.	8.	80.	Available	AL.01.5
BLL13035	Black bar	35.	8.	80.	Available	AL.01.7
BLL13040	Black bar	40.	8.	85.	Available	AL.01.8
BLL13045	Black bar	45.	8.	95.	Available	AL.01.9
BRL13020	Bright bar	20.	8.	130.	Available	AL.02.4
BRL13025	Bright bar	25.	8.	135.	Available	AL.02.5

STOCK INVOICE   DATE: 01/24/89

* * * * * * * * * * * * *

Institute of Manufacturing Engineering
Technical University of Denmark
DK-2900 Lyngby, Denmark

This STOCK INVOICE is used for the RWA MATERIALS DATABASE
of the expert process planning system XPLAN-R

Property of stock: Medium carbon steel 45#

Codes	Features	Dia. (mm)	Length (m)	Cost/m (DKK)	Stock state	Location
HXM45040	Hexagonal bar	40.	10.	285.	Available	MC.04.8
HXM45045	Hexagonal bar	45.	10.	285.	-0-	MC.04.9
SQM45020	Squaredsection	20.	5.	250.	Available	MC.05.3
SQM45025	Squaredsection	25.	5.	280.	-0-	MC.05.4
SQM45030	Squaredsection	30.	5.	285.	Available	MC.05.5
SQM45035	Squaredsection	35.	-0-	-0-	-0-	-0-
SQM45040	Squaredsection	40.	5.	320.	-0-	MC.05.6
SQM45045	Squaredsection	45.	-0-	-0-	-0-	-0-
BLM45020	Black bar	20.	8.	100.	Available	MC.01.4
BLM45025	Black bar	25.	8.	100.	Available	MC.01.5
BLM45030	Black bar	30.	8.5	110.	Available	MC.01.6
BLM45035	Black bar	35.	9.	110.	Available	MC.01.7
BLM45040	Black bar	40.	10.	120.	Available	MC.01.8
BLM45045	Black bar	45.	10.	120.	Available	MC.01.9
BRM45020	Bright bar	20.	10.	150.	Available	MC.02.4
BRM45025	Bright bar	25.	10.	160.	Available	MC.02.5
BRM45030	Bright bar	30.	10.	165.	Available	MC.02.6
BRM45035	Bright bar	35.	10.	170.	Available	MC.02.7
BRM45040	Bright bar	40.	10.	175.	Available	MC.02.8
BRM45045	Bright bar	45.	10.	180.	-0-	MC.02.9
HXM45020	Hexagonal bar	20.	8.	250.	Available	MC.03.4
HXM45025	Hexagonal bar	25.	8.	255.	-0-	MC.03.5
HXM45030	Hexagonal bar	30.	8.	265.	Available	MC.03.6
HXM45035	Hexagonal bar	35.	10.	275.	-0-	MC.04.7

The block diagrams of the operation sequences

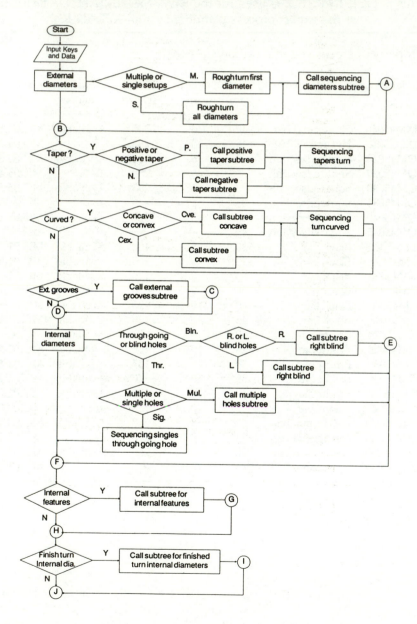

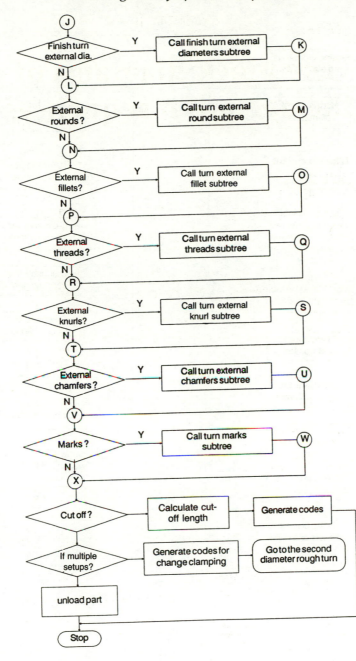

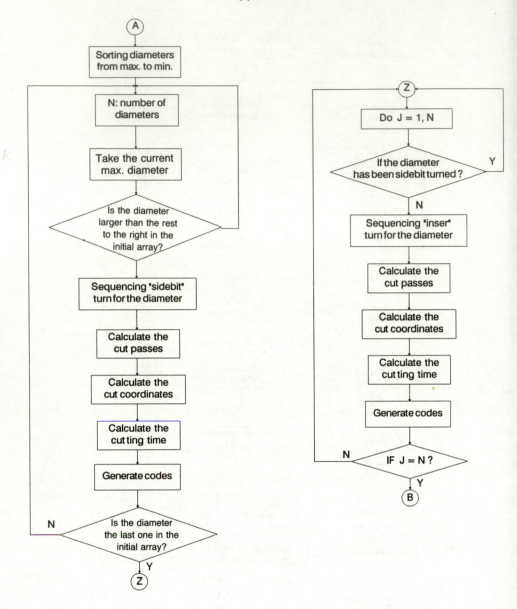

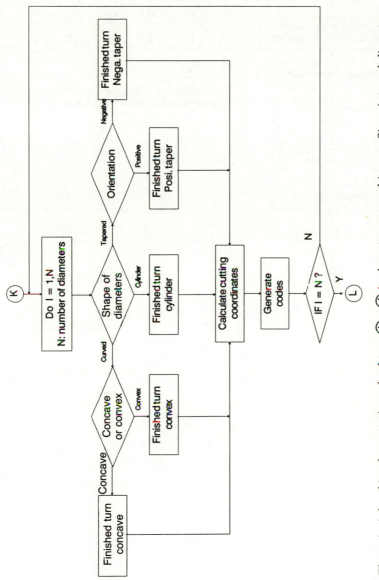

The principle of implementation of subtree (I)→(J) is almost same as this one. Since internal diameters should not be curved, the left loop of this diagram does not appear in subtree (I)→(J).

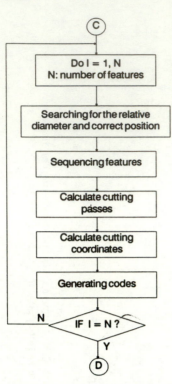

The following subtrees:

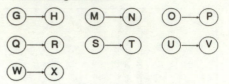

are implemented using the same principle as this one. Only the specific feature is involved.

Indication of codes
for tool selection

1. Codes for Process Sequencing: ⟨XX⟩
   RS : Rough sidebit turn
   RI : Rough insert turn
   FT : Finished turn
   GS : Groove squared
   GR : Groove round
   GH: Groove shaped
   TH : Threads turn
   CF : Chamfer degree 45
   CS : Chamfer degree 60
   FL : Fillet turn
   RD : Round turn
   FS : Face section
   CO: Cut off

   ⋮

2. Codes for Cutting Surface Shape: ⟨0⟩
   1: External surface
   5: Internal surface
   9: Section surface
3. Codes for Cutting Dimension Group: ⟨0⟩
   The dimensions of turning diameters can be grouped into several groups, according to the swing of each machine. For example, for the CINTURN 130 CNC lathe, the max. cutting diameter is 130 mm. Then this range can be grouped into 2 groups, i.e. (1) 5–30 mm and (2) 30–130 mm. Some machines' cutting dimensions only need to be grouped into a single group.
4. According to the initial tool materials, codes for tool materials can be any two alphabetic characters or ISO codes.

5. Codes for Tools' Geometrical Angles
   degree: 45, 75, 60, 90, etc.
6. Codes for Workpiece Materials

Corresponding to the material which is defined in the materials classification tree.

Screen display for running examples

X P L A N – R
Process Planning System
version 2.0

(C) Copyright by
Institute for Product Development
Laboratory of Process and Production Engineering
Building 425, Technical University of Denmark

Enter name
›› H.C. Zhang

Enter Password
››

XPLAN–R Main menu
1. Generative Process Planning
2. Print last generated Plan
3. Plan with alternative machines
4. Options
5. Database management
99. EXIT
››1

Enter ID-number ››ZT-SHA

Enter Description: Shaft group parts
››

Main shape
*   1 – Rotational
2 – Prismatic
3 – Other mainshape
**›
– Enter batch size
Value = 100

>>

Material
* 1 – Metal
2 – Plast
3 – Other
**>

Metal
* 1 – Steel
2 – Aluminum
3 – Brass MS58
4 – Inconel
5 – Berylco
6 – Other
**>

Steel
* 1 – Automate steel
2 – Stainless 43
**>

Condition
* 1 – Bar
2 – Semi product
3 – Other
**>

Bar
* 1 – Rod
2 – Hexagon
**>

Finest surface
1 – Finest surface roughness GE 3.2
* 2 – Finest surface roughness LT 3.2
**>

Closest tolerance
1 – Closest tolerance GT 0.01
* 2 – Closest tolerance LE 0.01
**>

Surface treatment
1 – Galvanize
* 2 – No surface treatment
**>

Rotational mainshape
* 1 – Single axis
2 – Multiple axis
**,

Single axis
* 1– External features.
2 – Internal features.
**,
– Enter number of external diameters.
Value = 7
››

Shape of dia. no. 1
* 1 – Straight
2 – Curved
**,

Straight
* 1 – Cylinder
2 – Taper
**,
– Enter diameter no. 1
Value = 12.00
›› – Enter length of dia. no. 1
Value = 22.00
››

Surface roughness for diameter no. 1
1 – Surface roughness GE 3.2
* 2 – Surface roughness LT 3.2 and GE 1.6
3 – Surface roughness LT 1.6
**,

Tolerance for diameter no. 1
1 – Tolerance GT 0.01 mm
* 2 – Tolerance LE 0.01 mm
**,

+ External features for dia. no. 1
Type ⟨XI⟩ to read information, if you need.
* 1 – Chamfer
2 – Round
3 – Fillet
* 4 – Thread
5 – Knurl
* 6 – Special features
7 – None
**,

Chamfer position on dia. no. 1
* 1 – Left chamfer
  2 – Right chamfer
**,

    – Enter LEFT chamfer angle.
      Value = 45.00
››

    – Enter LEFT chamfer width (mm) (X-Axis).
      Value = 1.00
››

Thread position on dia. no. 1
* 1 – UNC thread
  2 – UNF thread
**,

    – Enter UNC thread length
      Value = 20.00
››

    – Enter DELTA position of UNC thread
      Value = 0.00
››

Special features for dia. no. 1
* 1 – Grooves
  2 – Keyways
  3 – Slots
  4 – Gear
  5 – Marking
  6 – None
**,

+ Groove position on dia. no. 1
Type <XI> to read graphic illustration for grooves.
* 1 – Square groove
  2 – Round groove
  3 – Shape groove
**,

    – Enter square gro. width (mm) (X-Axis).
      Value = 2.00
›› – Enter square gro. depth (mm) (Y-Axis).
      Value = 1.50
›› – Enter DLETA position of gro. (mm) (X-Axis).
      Value = 20.00
››

Shape of dia. no. 2
*   1 – Straight
    2 – Curved
**,

Straight
*   1 – Cylinder
    2 – Taper
**,
   – Enter diameter no. 2
      Value = 30.00
  ››
   – Enter length of dia. no. 2
      Value = 23.00
››

Surface roughness for diameter no. 2
1 – Surface roughness GE 3.2
2 – Surface roughness LT 3.2 and GE 1.6
*   3 – Surface roughness LT 1.6
**,

Tolerance for diameter no. 2
1 – Tolerance GT 0.01 mm
*   2 – Tolerance LE 0.01 mm
**,

   + External features for dia. no. 2
    Type ⟨XI⟩ to read information, if you need.
    1 – Chamfer
    2 – Round
    3 – Fillet
    4 – Thread
    5 – Knurl
*   6 – Special features
    7 – None
**,

Special features for dia. no. 2
*   1 – Grooves
    2 – Keyways
    3 – Slots
    4 – Gear
    5 – Marking
    6 – None
**,

+ Groove position on dia. no. 2
  Type ⟨XI⟩ to read graphic illustration for grooves.
* 1 – Square groove
  2 – Round groove
  3 – Shape groove
**⟩

  – Enter square gro. width (mm) (X-Axis).
    Value = 2.50
⟩⟩ – Enter square gro. depth (mm) (Y-Axis).
    Value = 1.50
⟩⟩ – Enter DLETA position of gro. (mm) (X-Axis).
    Value = 20.50

⟩⟩

Shape of dia. no. 3
* 1 – Straight
  2 – Curved
**⟩

Straight
* 1 – Cylinder
  2 – Taper
**⟩

  – Enter diameter no. 3
    Value = 95.00

⟩⟩

  – Enter length of dia. no. 3
    Value = 24.00
**⟩

Surface roughness for diameter no. 3
1 – Surface roughness GE 3.2
* 2 – Surface roughness LT 3.2 and GE 1.6
  3 – Surface roughness LT 1.6
**⟩

Tolerance for diameter no. 3
1 – Tolerance GT 0.01 mm
* 2 – Tolerance LE 0.01 mm
**⟩

+ External features for dia. no. 3
  Type ⟨XI⟩ to read information, if you need.
* 1 – Chamfer
  2 – Round
  3 – Fillet
  4 – Thread

```
    5 – Kunrl
*   6 – Special features
    7 – None
**, XI
```

> The features of each diameter are divided into two groups. The first group contains general features, as you have seen on the screen. The second group contains special features, which are KEYWAYS, SLOTS, THREADS, GEARS, MARKING, KNURLS You can select any special features by typing 5.

```
    -RETURN- to Continue
 ,,
```

```
  + External features for dia. no. 3
    Type ⟨XI⟩ to read information, if you need.
*   1 – Chamfer
    2 – Round
    3 – Fillet
    4 – Thread
    5 – Knurl
*   6 – Special features
    7 – None
**,
```

```
    Chamfer position on dia. no. 3
*   1 – Left chamfer
*   2 – Right chamfer
**,
    – Enter LEFT chamfer angle.
       Value = 45.00
 ,,
    – Enter LEFT chamfer width (mm) (X-Axis).
       Value = 1.50
 ,,
    – Enter RIGHT chamfer Angle.
       Value = 45.00
 ,,
    – Enter RIGHT chamfer width (mm) (X-Axis).
       Value = 1.50
 ,,
```

Special features for dia. no. 3
1 – Grooves
2 – Keyways
3 – Slots
* 4 – Gear
5 – Marking
6 – None
**›

Shape of dia. no. 4
* 1 – Straight
2 – Curved
**›

Straight
* 1 – Cylinder
2 – Taper
**›

– Enter diameter no. 4
Value = 34.00
››

– Enter length of dia. no. 4
Value = 30.00
››

Surface roughness for diameter no. 4
1 – Surface roughness GE 3.2
* 2 – Surface roughness LT 3.2 and GE 1.6
3 – Surface roughness LT 1.6
**›

Tolerance for diameter no. 4
1 – Tolerance GT 0.01 mm
* 2 – Tolerance LE 0.01 mm
**›

+ External features for dia. no. 4
Type ⟨XI⟩ to read information, if you need.
1 – Chamfer
2 – Round
* 3 – Fillet
4 – Thread
5 – Knurl
6 – Special features
7 – None
**›

Fillet position on dia. no. 4
* 1 – Left fillet
2 – Right fillet
**,
– Enter LEFT fillet ridius.
Value = 5.00
››

Shape of dia. no. 5
* 1 – Straight
2 – Curved
**,

Straight
* 1 – Cylinder
2 – Taper
**,
– Enter diameter no. 5
Value = 25.00
››
– Enter length of dia. no. 5
Value = 25.00
››

Surface roughness for diameter no. 5
1 – Surface roughness GE 3.2
* 2 – Surface roughness LT 3.2 and GE 1.6
3 – Surface roughness LT 1.6
**,

Tolerance for diameter no. 5
1 – Tolerance GT 0.01 mm
* 2 – Tolerance LE 0.01 mm
**,

+ External features for dia. no. 5
Type ⟨XI⟩ to read information, if you need.
1 – Chamfer
2 – Round
3 – Fillet
4 – Thread
5 – Knurl
* 6 – Special features
7 – None
**,

Special features for dia. no. 5
1 – Grooves

    2 – Keyways
    3 – Slots
    4 – Gear
*   5 – Marking
    6 – None
**⟩

Marking position on dia. no. 5
    1 – Marking parallel axis
*   2 – Marking vertical axis
    3 – Hand marking
**⟩

    – Enter marking.
        Value = MADE IN DENMARK
  ⟩⟩

Shape of dia. no. 6
*   1 – Straight
    2 – Curved
**⟩

Straight
*   1 – Cylinder
    2 – Taper
**⟩

    – Enter diameter no. 6
        Value = 50.00
  ⟩⟩

    – Enter length of dia. no. 6
        Value = 32.00
  ⟩⟩

Surface roughness for diameter no. 6
    1 – Surface roughness GE 3.2
*   2 – Surface roughness LT 3.2 and GE 1.6
    3 – Surface roughness LT 1.6
**⟩

Tolerance for diameter no. 6
    1 – Tolerance GT 0.01 mm
*   2 – Tolerance LE 0.01 mm
**⟩

  + External features for dia. no. 6
    Type ⟨XI⟩ to read information, if you need.
    1 – Chamfer
    2 – Round
    3 – Fillet

4 – Thread
5 – Knurl
*   6 – Special features
7 – None
**,

Special features for dia. no. 6
1 – Grooves
*   2 – Keyways
3 – Slots
4 – Gear
5 – Marking
6 – None
**,

+ Keyway position on dia. no. 6
Type ⟨XI⟩ to read graphic illustration for keyways.
*   1 – Keyway
2 – Woodruff key
3 – Spline
**,

– Enter keyway length (mm) (X-Axis).
   Value = 22.00
››

– Enter keyway depth (mm) (Y-Axis).
   Value = 5.00
››

– Enter keyway width (mm) (Z-Axis).
   Value = 8.00
››

– Enter DELTA position of keyway (mm) (X-Axis).
   Value = 5.00
››

Shape of dia. no. 7
*   1 – Straight
2 – Curved
**,

Straight
*   1 – Cylinder
2 – Taper
**,

– Enter diameter no. 7
  Value = 35.00
››

– Enter length of dia. no. 7

Value = 30.00
››

Surface roughness for diameter no. 7
1 – Surface roughness GE 3.2
2 – Surface roughness LT 3.2 and GE 1.6
*   3 – Surface roughness LT 1.6
**›

Tolerance for diameter no. 7
1 – Tolerance GT 0.01 mm
*   2 – Tolerance LE 0.01 mm
**›

+ External features for dia. no. 7
Type ⟨XI⟩ to read information, if you need.
*   1 – Chamfer
    2 – Round
    3 – Fillet
    4 – Thread
    5 – Knurl
*   6 – Special features
    7 – None
**›

Chamfer position on dia. no. 7
1 – Left chamfer
*   2 – Right chamfer
**›

– Enter RIGHT chamfer Angle.
Value = 45.00
››

– Enter RIGHT with chamfer width (mm) (X-Axis).
Value = 1.50
››

Special features for dia. no. 7
*   1 – Grooves
    2 – Keyways
    3 – Slots
    4 – Gear
    5 – Marking
    6 – None
**›

+ Groove position on dia. no. 7
Type ⟨XI⟩ to read graphic illustration for grooves.
*   1 – Square groove

2 – Round groove
3 – Shape groove
**›

– Enter square gro. width (mm) (X-Axis).
  Value = 2.50
›

– Enter square gro. depth (mm) (Y-Axis).
  Value = 1.50
›

– Enter DLETA position of gro. (mm) (X-Axis).
  Value = 0.00
›

Final inspection
1 – Inspection
2 – Material testing
* 3 – No inspection
**›

Choose Option:
1 – Review Choices
* 2 – Continue
==›

   *** OLD Entry: ZT-SHA   KA

Choose Option:
* 1 – Store ID #
  2 – Change ID Name
  3 – No Store
==› 3

*** The following processes were selected ***
   process #1: Process: Raw Material Storage
   process #2: Process: First Turning
   process #3: Process: Second Turning
   process #4: Process: Finish Turning
   process #5: Process: Milling

*** The following machines were found for ***
   process: #1: Process: Raw Material Storage

   UA Raw Materials

*** The following machines were found for ***
   process #2: Process: First Turning

   LK TU09
   LC TU10

*** The following machines were found for ***
process #3: Process: Second Turning

LB TU08
LL TU05

*** The following machines were found for ***
process #4: Process: Finish Turning

LM TU04
LG TU06

*** The following machines were found for ***
process #5: Process: Milling

FA ML01
FC ML04
FB ML03

*** SORTING OF MACHINES ***

*** OPERATION DECISION ***

*** Operation decision for ***
machine #1: UA Raw Materials

*** Operation decision for ***
machine #2: LC TU10

*** Operation decision for ***
machine #3: LB TU08

*** Operation decision for ***
machine #4: LG TU06

*** Operation decision for ***
machine #5: FA ML01

*** GENERATION OF PLAN ***

Enter output destination:
*   1. Screen
2. Printer
3. User selected file
  ››3
Enter filename and extension (ex. PLAN. OUT)
  ›› ZT-SHA. OUT
File already exists, overwrite?

››Y
Writing plan to existing file ZT-SHA. OUT
XPLAN-R Main menu
1. Generative Process Planning
2. Print last generated Plan
3. Plan with alternative machines
4. Options
5. Database management
99. EXIT
››

Exercises

CHAPTER 1

1.1 What is the keyword used to distinguish manufacturing industry from process industry? Can you point out some examples of both?

1.2 A tobacco company can be considered a _____
   a. *manufacturing industry*
   b. *process industry*

1.3 Give a brief discussion of different production types.

1.4 Define the term 'manufacturing systems' and explain the reasons for studying manufacturing systems.

1.5 Describe the various types of manufacturing systems.

1.6 What is the difference between Dedicated Manufacturing System (DMS) and Flexible Manufacturing System (FMS)?

1.7 What is a manufacturing process? Give a brief discussion of different material manufacturing processes.

1.8 What is manufacturing planning? Discuss the aspects of the major planning activities involved in manufacturing.

CHAPTER 2

2.1 Define manufacturing process planning, and explain the role played by process planning in manufacturing industry.

2.2 Define the following terms:
   a. Process
   b. Operation
   c. Cut.

2.3   What essential information is required for process planning?

2.4   Explain the approaches used for process planning.

2.5   Give the advantages and disadvantages of using the manual process planning approach.

2.6   What are the usages of the process plans?

2.7   Briefly describe the steps of process planning.

2.8   What are datums? Describe the various types of datum.

2.9   Why is selecting a datum important? Define the terms *bright datum*, *black datum*, and *auxiliary datum*.

2.10   What is meant by synchronous datum? Discuss briefly.

2.11   Define the terms *datum unification, datum coincidence,* and *process concentration*.

2.12   Discuss the criteria that must be considered while sequencing the operations.

2.13   Prepare a process plan for the rotational part shown in Figure 2.9.

2.14   Prepare a process plan for the prismatic part shown in Figure 2.12.

CHAPTER 3

3.1   What is CIM? What is the difference between CIM and Automation?

3.2   What are the basic steps of engineering designs?

3.3   Describe the different types of geometrical model used in CAD systems.

3.4   What are CAD and CAM?

3.5   Describe the manufacturing functions that are currently addressed by CAM.

3.6   What is Numerical Control (NC)?

3.7   What is CNC? What is DNC? What is the difference between CNC and DNC?

3.8   Why is there a need for a computer monitoring and diagnostic process in a manufacturing industry?

3.9   What is a database system? Discuss different types of database systems.

3.10   What are the three levels of database architecture? Explain.

3.11   What is a distributed system? Indicate its advantages.

3.12   What is the difference between the databases used in engineering and those used in business.

3.13   Explain the role of a manufacturing database in process planning.

3.14   What are Expert Systems? What are the basic components of Expert Systems? Describe their application.

3.15   What are the three general types of knowledge?

3.16   Describe different knowledge representation approaches.

3.17   What is an inference engine? Describe different inference approaches. What are forward chaining and backward chaining?

3.18   Describe the reasons for using expert systems in manufacturing process planning.

3.19   What is a semantic network?

3.20   What is a frame-based system?

3.21   What is a hybrid approach?

CHAPTER 4

4.1   What is Group Technology (GT)? What is a part family?

4.2   What is classification and coding?

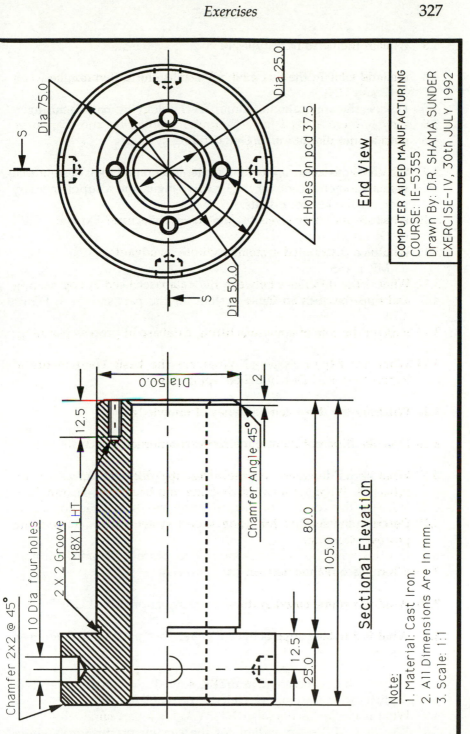

End View

Dia 75.0

Dia 25.0

Dia 50.0

4 Holes Qn pcd 37.5

COMPUTER AIDED MANUFACTURING
COURSE: IE-5355
Drawn By: D.R. SHAMA SUNDER
EXERCISE-IV, 30th JULY 1992

Chamfer 2x2 @ 45°

10 Dia four holes

2 X 2 Groove

M8X1 LHT

Dia 50.0

12.5

2

Chamfer Angle 45°

80.0

105.0

12.5

25.0

Sectional Elevation

Note:
1. Material: Cast Iron.
2. All Dimensions Are In mm.
3. Scale: 1:1

Figure 4.14

4.3 Explain the three basic types of coding systems.

4.4 List and explain the five methods used to form part families.

4.5 Discuss the application of Group Technology in manufacturing.

4.6 Give a brief discussion of Opitz coding system.

4.7 Which of the following coding systems is a nine digit system that uses a geometric code for the first five digits and a supplementary code for the last four digits.
    a. KK-3
    b. DCLASS
    c. Opitz
    d. MICLASS

4.8 Develop the code in Opitz system for the part shown in Figure 4.14.

CHAPTER 6

6.1 Define computer aided process planning.

6.2 Discuss the goals of computerized process planning systems.

6.3 Give a brief discussion of the variant approach of CAPP.

6.4 What is the concept of 'master part' and how is it applied in process planning?

6.5 Discuss how the variant approach to process planning is more advantageous than the manual approach.

6.6 Give a brief discussion of the generative approach of CAPP.

6.7 What is the difference between 'generative' and 'semi-generative' systems?

6.8 Why is the generative approach called the 'highest level of sophistication in computer aided process planning'?

6.9 Why is CAPP so important for the current production environment?

6.10 What are the CAPP benefits in manufacturing and what are the new developments in CAPP systems today?

6.11 What are the 3 different aspects of the new generation of CAPP compared with traditional CAPP systems?

Index